Current Research in Thermal Conductivity

Edited by Roberto Palma Guerrero

Published in London, United Kingdom

Current Research in Thermal Conductivity
http://dx.doi.org/10.5772/intechopen.1003411
Edited by Roberto Palma Guerrero

Contributors
Afshin Kouhkord, Aondoyila Kuhe, Dianta Ginting, Elif Begum Elcioglu, Faridoddin Hassani, Fatemeh Rezaei, Jong-Soo Rhyee, José Miguel Palomino Cobo, Keyvan Homayouni, María Esther Puertas García, Ndah Abdulrahman Alpha, Rafael Gallego Sevilla, Ramin Farzadi, Roberto Palma Guerrero, Soroush Javadipour

First published in London, United Kingdom, 2025 by IntechOpen
IntechOpen is the global imprint of INTECHOPEN LIMITED, registered in England and Wales, registration number: 11086078, 167-169 Great Portland Street, London, W1W 5PF, United Kingdom

For EU product safety concerns: IN TECH d.o.o., Prolaz Marije Krucifikse Kozulić 3, 51000 Rijeka, Croatia, info@intechopen.com or visit our website at intechopen.com.

British Library Cataloguing-in-Publication Data
A catalogue record for this book is available from the British Library

Current Research in Thermal Conductivity
Edited by Roberto Palma Guerrero
p. cm.
Print ISBN 978-1-83769-757-1
Online ISBN 978-1-83769-756-4
eBook (PDF) ISBN 978-1-83769-758-8

If disposing of this product, please recycle the paper responsibly.

Meet the editor

Roberto Palma earned his Physics BSc (2005), MSc (2006) and Ph.D. (2012) from the University of Granada (Spain) receiving two awards for his thesis. He worked on High-Energy Physics at the Universitat Politècnica de València (2010-2012) and was an Assistant Professor at Universitat Jaume I (2013-2019). Since 2019, he has been an Associate Professor at the University of Granada. His research focuses on thermodynamically consistent formulations for nonlinear finite elements, modelling of smart materials, soft tissues, tumour growth, and plasticity in irradiated materials. He has published 34 journal articles with an h-index of 19 and has participated in 18 publicly funded research projects and three industry collaborations.

Contents

Preface

Thermal conductivity plays a crucial role in the fundamental processes that govern heat transfer and entropy, deeply rooted in the second law of thermodynamics. As temperature gradients arise in materials, they generate heat fluxes dictated by a key material property: thermal conductivity. Understanding this property is essential not only for interpreting thermodynamic principles but also for addressing modern challenges in science and engineering.

In recent years, thermal conductivity has gained renewed attention due to its significance in emerging technologies and materials. From nanofluids and nanocomposites to advanced nuclear fusion and bioheat applications, innovations in controlling and enhancing thermal conductivity are expanding the boundaries of scientific and technological possibilities. These advancements are reshaping industries, with potential applications ranging from energy efficiency improvements to breakthroughs in medicine and materials science.

This book emerges from the need to bridge the gap between theoretical understanding and practical application of thermal conductivity in cutting-edge fields. It combines recent developments and discoveries, demonstrating how a deep understanding of thermal properties can lead to transformative innovations. Through this work, I aim to provide a comprehensive overview of how thermal conductivity impacts a broad range of applications, offering readers a solid foundation for advancing their own research or industrial pursuits.

For researchers and engineers, this book serves as a guide to navigating the key mechanisms that govern thermal processes while also highlighting the challenges that must be overcome to push the boundaries of modern science. Whether you are working in academia or industry, the knowledge presented here will equip you with the tools to explore the next generation of thermal management solutions.

I invite you to journey through the complexities of thermal conductivity and discover how this essential property is opening new doors in the fields of nanotechnology, energy systems, and beyond.

Roberto Palma
Department of Structural Mechanics,
University of Granada,
Granada, Spain

Chapter 1

Optimizing Thermal Conductivity in PbTe: Nanocomposite and Alloy Approaches for Low Thermal Conductivity

Dianta Ginting and Jong-Soo Rhyee

Abstract

PbTe, a prominent thermoelectric material within the intermediate temperature range (500 K to 950 K), has displayed noteworthy potential due to its cubic rock salt crystal structure and narrow band gap of 0.32 eV. This investigation explores the quaternary system $(PbTe)_{0.95-x}$-$(PbSe)_x$-$(PbS)_{0.05}$ with x = 0, 0.05, 0.10, 0.15, 0.20, 0.35, and 0.95, along with a consistent Na dopant concentration of 1 at%. The findings indicate that the inclusion of PbSe and PbS significantly reduces the lattice thermal conductivity, with the lowest value observed in $(PbTe)_{0.75}$-$(PbSe)_{0.20}$-$(PbS)_{0.05}$, achieving 0.42 Wm^{-1} K^{-1}, nearing the glass limit for bulk PbTe. A detailed examination using transmission electron microscopy (TEM) identifies nanostructuring as a critical mechanism for the observed reduction in thermal conductivity. The study's outcomes highlight the crucial role of nanostructured precipitates in enhancing phonon scattering, thereby reducing thermal conductivity.

Keywords: thermoelectric materials, PbTe, nanocomposites, alloying, thermal conductivity, lattice thermal conductivity, nanostructuring

1. Introduction

Amidst mounting energy and environmental crises worldwide, the imperative for effective energy utilization and the advancement of eco-friendly energy sources has become paramount. Thermoelectric materials, adept at transforming heat into electricity, emerge as a promising solution. They hold the potential to reshape cooling technology, obviating the reliance on traditional refrigerants. Nonetheless, their current low efficiency presents a formidable challenge, emphasizing the urgent requirement for continued progress to expedite their commercial viability [1]. The efficacy of thermoelectric materials is delineated by the dimensionless ZT figure-of-merit, expressed as $ZT = S^2\sigma T/\kappa$, where S represents the Seebeck coefficient, σ symbolizes the electrical conductivity, T denotes the absolute temperature, and κ signifies the thermal conductivity. The denominator $S^2\sigma$ is commonly referred to as the power factor. Enhancing material performance necessitates the identification of substances characterized by elevated Seebeck coefficient and electrical conductivity coupled with diminished thermal conductivity. However, these properties are interconnected, thereby constraining overall performance. Notably, augmenting the Seebeck coefficient tends to reduce electrical conductivity, with the latter being directly proportional to thermal conductivity as per the Wiedemann-Franz law, making independent parameter manipulation highly challenging. Many research endeavors have been pursued over recent decades, the pace of commercialization still is sluggish [2].

Lead tellurium (PbTe) stands out as one of the most efficient thermoelectric materials within the "intermediate temperature range," typically spanning from 500 to 950 K [3–5]. Exhibiting a cubic rock salt (halite) crystal structure within the Fm-3 m space group, this material is characterized by its narrow band gap of 0.32 eV. PbTe displays versatility in conductivity, readily shifting between N-type and P-type with the introduction of common dopants. Halogens such as $PbCl_2$, $PbBr_2$, and PbI_2 act as effective N-type dopants, fostering the formation of donor centers. Similarly, dopants like Bi_2Te_3, $TaTe_2$, and $MnTe_2$, when substituting for lead, induce vacancies in Pb sites. Conversely, for P-type conductivity, dopants, like Na_2Te, K_2Te, and Ag_2Te, substituting for tellurium, create vacant Te sites. Moreover, PbTe exhibits exceptional thermal stability, boasting a high melting point of 1190 K, rendering it suitable for thermoelectric applications at elevated temperatures up to 900 K [3]. Various efforts have been made to substitute Te with Se and S to develop PbSe [6, 7], PbS [8], or binary systems like PbTe-PbSe [9–11], PbTe-PbS [12–14], and PbSe-PbS [15, 16].

In the PbTe-PbSe system, the superior thermoelectric performance is predominantly attributed to band engineering. Band engineering in PbTe-PbSe is conducive to a high Seebeck coefficient, while achieving low resistivity necessitates doping [10, 17]. It is posited that the low lattice thermal conductivity in binary PbTe-PbSe systems is primarily caused by point defects resulting from a mixture of Te/Se in the rock salt structure and the presence of doping. Conversely, the exceptional thermoelectric properties in PbTe-PbS binary systems can be ascribed to a process called "nanostructuring," which involves the creation of nanoscale structures within the material. Nanostructuring in PbTe-PbS systems occurs through bulk phase separation, whether by nucleation or spinodal decomposition, contingent upon the relative phase fractions [12, 18, 19].

Recently, a novel and captivating area of research has emerged, focusing on the complexity of quaternary systems such as PbTe-PbSe-PbS [20–22]. These systems, unlike their binary counterparts, exhibit superior thermoelectric performance. The unique and remarkable thermoelectric properties in quaternary PbTe-PbSe-PbS systems can be attributed to low lattice thermal conductivity. This low lattice thermal conductivity in PbTe-PbSe-PbS systems primarily arises from point defects due to triple disorder in the rock salt structure. However, the effect of nanostructuring on lattice thermal conductivity in quaternary PbTe-PbSe-PbS systems has not been observed in previous reports, making it a fascinating and unexplored area for further investigation.

Given the high thermoelectric performance of PbTe-PbSe and PbTe-PbS systems, this study investigates the quaternary system of $(PbTe)_{0.95\text{-}x}\text{-}(PbSe)_x\text{-}(PbS)_{0.05}$ (x = 0, 0.05, 0.10, 0.15, 0.20, 0.35, and 0.95). The initial examination of the thermoelectric performance of quaternary lead chalcogenide composites was conducted with a sodium (Na) dopant concentration of 1 at%, corresponding to $Pb_{0.98}Na_{0.02}Te_{0.95\text{-}x}Se_xS_{0.05}$. The PbS concentration was fixed at 5% in all samples to maintain consistency in the PbTe electronic band structure, which sulfur (S) solute atoms can alter, similar to the ternary $(PbTe)_{1\text{-}x}$-$(PbS)_x$ system. Changes in the electronic band structure are solely linked to the fraction of PbSe in the $(PbTe)_{0.95\text{-}x}$-$(PbSe)_x$-$(PbS)_{0.05}$ composition. The study revealed an exceptionally low lattice thermal conductivity of 0.42 Wm^{-1} K^{-1} for $(PbTe)_{0.75}$-$(PbSe)_{0.20}$-$(PbS)_{0.05}$, approaching the glass limit for bulk PbTe systems (0.36 Wm^{-1} K^{-1}). Notably, the low lattice thermal conductivity in $(PbTe)_{0.75}$-$(PbSe)_{0.20}$-$(PbS)_{0.05}$ was primarily attributed to nanostructuring. This study delves into the phenomenon of solid-state nucleation to create nanostructured precipitates in the quaternary system PbTe-PbSe-PbS.

2. Experimental details

2.1 Sample preparations

Polycrystalline PbSe and PbS samples were prepared through the synthesis process involving the mixing of high-purity Pb (99.999%), dried S (99.999%), and Se (99.999%) in vacuum-sealed quartz ampoules at a residual pressure of 10–4 Torr. These components were then reacted at 1150°C to yield high-purity starting materials. The subsequent development of Polycrystalline $Pb_{0.98}Na_{0.02}Te_{0.95\text{-}x}Se_xS_{0.05}$, also recognized as $(PbTe)_{0.95\text{-}x}$-$(PbSe)_x$-$(PbS)_{0.05}$ (x = 0, 0.05, 0.1, 0.15, 0.20, 0.35 and 0.95), doped with 1% Na, was accomplished by combining stoichiometric amounts of high purity PbSe, PbS, Pb, Te, and Na as the dopant. The total mass of 20 g was loaded into carbon-coated fused silica tubes, flame sealed under a vacuum of 10^{-4} Torr, and heated to 1373°C. Following homogenization at 1373 K for 10 hours, the samples were quenched in cold water and then annealed at 774°C for 72 hours. Subsequently, the ingots were manually ground into powder form. These powders were then pressed into dish-shaped forms (12.7 mm diameter) using a hot press with 40 MPa of axial pressure applied for 1 hour at 773 K under an argon atmosphere by hot press.

2.2 Sample characterization

The examination of microstructure was carried out using transmission electron microscopy (TEM) (LIBRA 200 MC, ZEISS, Germany). The TEM samples underwent thorough grinding, were evenly dispersed in alcohol through sonication, and were dried prior to examination under the microscope.

The calculation of thermal conductivity (κ) is derived from the equation $\kappa = \rho_s \bullet D_T \bullet C_p$, where ρ_s, D_T, and C_p represent the sample density, thermal diffusivity, and specific heat, respectively. Determination of sample density (ρs) was conducted using the Archimedes method based on measurements of sample weight and dimensions. The thermal diffusivity (D_T) was determined through the utilization of the laser flash method (LFA-447, NETZSCH, Germany). Estimation of specific heat (C_p) was measured by a physical property measurement system (PPMS, Dynacool 14 T, Quantum Design, U.S.A.) and extrapolated to high-temperature region.

2.3 Result and discussion

The graphical representation in **Figure 1a** illustrates the comprehensive thermal conductivity of various compositions. It is noteworthy that an increase in PbSe

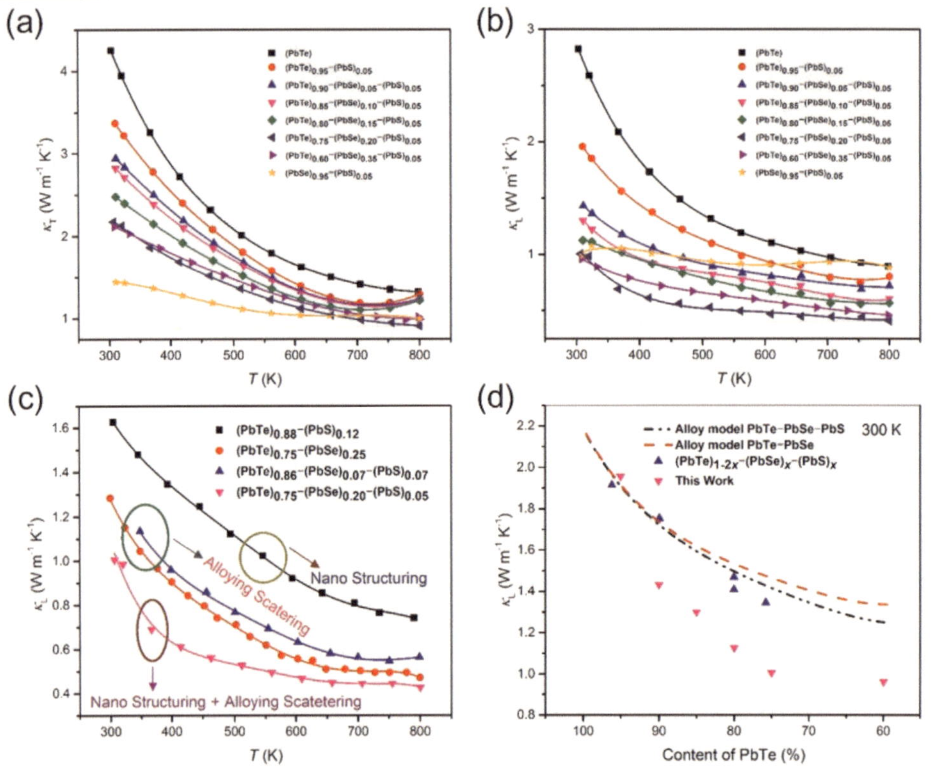

Figure 1.
Total thermal conductivity of $(PbTe)_{0.95\text{-}x}$-$(PbSe)_x$-$(PbS)_{0.05}$, (b) lattice thermal conductivity of $(PbTe)_{0.95\text{-}x}$-$(PbSe)_x$-$(PbS)_{0.05}$, (c) lattice thermal conductivity of $(PbTe)_{0.75}$-$(PbSe)_{0.20}$-$(PbS)_{0.05}$ compare with lattice thermal conductivity of $(PbTe)_{0.88}$-$(PbS)_{0.12}$, $(PbTe)_{0.75}$-$(PbSe)_{0.25}$ and $(PbTe)_{0.86}$-$(PbSe)_{0.07}$-$(PbS)_{0.07}$ (d) comparison of theoretical lattice thermal conductivity for (PbTe)-(PbSe) alloy and (PbTe)-(PbSe)-(PbS) alloy base in Ref. [20].

content led to a reduction in the overall thermal conductivity. A declining trend in total thermal conductivity is observed at $x \leq 0.15$ until 700 K, where a reversal in this trend begins, a phenomenon also evident in $(PbTe)_{0.9}$-$(PbS)_{0.1}$ doped with 3% Na [13]. Conversely, there is a notable decrease in total thermal conductivity as temperature rises for $x = 0.20$ and $x = 0.35$. The composition with $x = 0.20$ exhibited the lowest total thermal conductivity values, measuring 2.17 $Wm^{-1} K^{-1}$ at room temperature and 0.90 $Wm^{-1} K^{-1}$ at elevated temperatures. The incorporation of 5% PbS and 20% PbSe resulted in a 48% reduction in total thermal conductivity at room temperature and a 31.8% decrease at higher temperatures compared to pristine PbTe. The low total thermal conductivity for $(PbTe)_{0.95\text{-}x}$-$(PbSe)_x$-$(PbS)_{0.05}$ values imply the low lattice thermal conductivity. To fully visualize this effect, the lattice thermal conductivity was calculated by subtracting the value of the charge carrier thermal conductivity (κ_{el}) from the Wiedemann-Franz relation: $\kappa_{el} = LT/\rho$ from κtot such that $\kappa_{latt} = \kappa_{tot} - LT/\rho$, where ρ is the resistivity, T represents the temperature, while L signifies the Lorenz number, which is determined as a function of temperature within a parabolic band featuring acoustic phonon scattering. [21], L was calculated from the reduced Fermi energy, which is derived via the experimental Seebeck coefficient as described below:

The Fermi integrals $Fj(\eta)$ can be derived as follows:

$$F_j(\eta) = \int_0^{\infty} f\epsilon^j d\epsilon = \int_0^{\infty} \frac{\epsilon^j d\epsilon}{1 + \exp(\epsilon - \eta)} \tag{1}$$

where η is the reduced electrochemical potential, which is available from the temperature dependent Seebeck coefficient as given by:

$$S = \frac{k}{e}\left(\frac{2F_1(\eta)}{F_0(\eta)} - \eta\right) \tag{2}$$

Finally, the temperature-dependent Lorenz number can be calculated by η values that fit the Eq. (2):

$$L = \left(\frac{k}{e}\right)^2 \frac{3F_0(\eta)F_2(\eta) - 4F_1(\eta)^2}{F_0(\eta)^2} \tag{3}$$

Figure 1b shows the lattice thermal conductivity, which can be calculated with the L values for the parabolic band. Lattice thermal conductivity values strongly decrease with the increase of PbSe content until $x = 0.20$ thereafter, it starts to increase. The lowest lattice thermals conductivities are found in $(PbTe)_{0.6}$-$(PbSe)_{0.35}$-$(PbS)_{0.05}$ with 0.95 $Wm^{-1} K^{-1}$ at 300 K and in $(PbTe)_{0.75}$-$(PbSe)_{0.20}$-$(PbS)_{0.05}$ with 0.40 $Wm^{-1} K^{-1}$ at 800 K. The lattice thermal conductivity values are strongly reduced by 66% at 300 K and 55% at 800 K compared with lattice thermal conductivity in PbTe. The lowest lattice thermal conductivity in $(PbTe)_{0.75}$-$(PbSe)_{0.20}$-$(PbS)_{0.05}$ at high temperature is approaching the glass limit for the bulk PbTe system (0.36 $Wm^{-1} K^{-1}$) [23]. To more understand the role of (PbSe) and $(PbS)_{0.05}$ in reducing lattice thermal conductivity, We compare the lattice thermal conductivity in $(PbTe)_{0.75}$-$(PbSe)_{0.20}$-$(PbS)_{0.05}$ with lead chalcogenide systems such as nanostructure system $(PbTe)_{0.88}$-$(PbS)_{0.12}$ [12], $(PbTe)_{0.75}$-$(PbSe)_{0.25}$ [10], $(PbTe)_{0.84}$-$(PbSe)_{0.07}$-$(PbS)_{0.07}$ [20] system (**Figure 1c**). **Figure 1c** shows clearly that the lattice thermal conductivity in $(PbTe)_{0.75}$-$(PbSe)_{0.20}$-$(PbS)_{0.05}$ system is lower by 45.94, 14.89, and 28.57% compared to lattice thermal conductivity in $(PbTe)_{0.88}$-$(PbS)_{0.12}$, $(PbTe)_{0.75}$-$(PbSe)_{0.25}$, and $(PbTe)_{0.84}$-$(PbSe)_{0.07}$-$(PbS)_{0.07}$ system. Low lattice thermal conductivity in $(PbTe)_{0.88}$-$(PbS)_{0.12}$ system is due to nanostructuring [18, 19]. On the contrary, in $(PbTe)_{0.75}$-$(PbSe)_{0.25}$ system, low lattice thermal conductivity in $(PbTe)_{0.75}$-$(PbSe)_{0.25}$ has been explained on effective point defect scattering created by the Te/Se mixed occupation in the rock salt structure in $(PbTe)_{0.75}$-$(PbSe)_{0.25}$ system [10]. Low lattice thermal conductivity in $(PbTe)_{0.75}$-$(PbSe)_{0.25}$ system known as a solid solution with lacking nanostructuring. Similar to $(PbTe)_{0.75}$-$(PbSe)_{0.25}$, in $(PbTe)_{0.84}$-$(PbSe)_{0.07}$- $(PbS)_{0.07}$ system found that nanostructuing was not occurred. The low lattice thermal conductivity in $(PbTe)_{0.84}$-$(PbSe)_{0.07}$-$(PbS)_{0.07}$ seems to arise by multiple types of point defects such as Te/S, Te/Se, and Se/S in $(PbTe)_{0.84}$-$(PbSe)_{0.07}$-$(PbS)_{0.07}$ rather than nanostructring [20]. At high temperatures, the lowest lattice thermal conductivity in $(PbTe)_{0.75}$-$(PbSe)_{0.20}$-$(PbS)_{0.05}$ approaches the glass limit for the bulk PbTe system, which is reported as 0.36 $Wm^{-1} K^{-1}$ [23].

To gain deeper insights into the influence of (PbSe) and $(PbS)_{0.05}$ on reducing lattice thermal conductivity, we compare the lattice thermal conductivity across different lead chalcogenide systems, including the nanostructure system $(PbTe)_{0.88}$-$(PbS)_{0.12}$ [12], $(PbTe)_{0.75}$-$(PbSe)_{0.25}$ [10], and $(PbTe)_{0.84}$-$(PbSe)_{0.07}$-$(PbS)_{0.07}$ [20] (**Figure 1c**).

The data depicted in **Figure 1c** highlights a decrease of 45.94, 14.89, and 28.57% in the lattice thermal conductivity of the $(PbTe)_{0.75}$-$(PbSe)_{0.20}$-$(PbS)_{0.05}$ system compared to the $(PbTe)_{0.88}$-$(PbS)_{0.12}$, $(PbTe)_{0.75}$-$(PbSe)_{0.25}$, and $(PbTe)_{0.84}$-$(PbSe)_{0.07}$-$(PbS)_{0.07}$ systems, respectively.

The reduced lattice thermal conductivity in the $(PbTe)_{0.88}$-$(PbS)_{0.12}$ system is attributed to nanostructuring. Conversely, in the $(PbTe)_{0.75}$-$(PbSe)_{0.25}$ system, the low lattice thermal conductivity is explained by effective point defect scattering resulting from Te/Se mixed occupation in the rock salt structure [10]. This system is recognized as a solid solution lacking nanostructuring. Similarly, in the $(PbTe)_{0.84}$-$(PbSe)_{0.07}$-$(PbS)_{0.07}$ system, nanostructuring is absent. Here, the diminished lattice thermal conductivity appears to be induced by various types of point defects such as Te/S, Te/Se, and Se/S, rather than nanostructuring [20].

In examining the low lattice thermal conductivity within the $(PbTe)_{0.75}$-$(PbSe)_{0.20}$-$(PbS)_{0.05}$ system, several potential mechanisms are under consideration. Point defects, observed in the $(PbTe)_{0.75}$-$(PbSe)_{0.25}$ and $(PbTe)_{0.84}$-$(PbSe)_{0.07}$-$(PbS)_{0.07}$ systems, are one such mechanism. Additionally, significant phonon scattering attributed to nanostructuring also contributes to the reduced lattice thermal conductivity. To further understand these mechanisms in the context of $PbTe)_{0.84}$-$(PbSe)_{0.07}$-$(PbS)_{0.07}$, a lattice thermal conductivity model for PbTe-PbSe and PbTe-PbSe-PbS, based on the Klemens approach [20, 24], is employed:

$$\kappa_{latt,alloy} = \kappa_{latt,pure} \frac{tan^{-1}(u)}{u}, u^2 = \pi \frac{\Theta_D \Omega}{2\hbar v^2} \kappa_{latt,pure} \Gamma \tag{4}$$

where Θ_D is the Debye temperature, Ω is the molar volume, υ is the velocity of sound, and Γ is a disorder scaling parameter that depends on mass and strain field fluctuations ($\Delta m/m$ and $\Delta\alpha/\alpha$).

Abeles extended the formula by Klemens, taking into account not only Umklapp and point defect scatterings but also phonon scattering by a normal process [20, 25]:

$$\kappa_{latt,alloy} = k_{latt,pure} \left(\frac{1}{1+\frac{5a}{9}}\right) \left[\frac{tan^{-1}u}{u} + \frac{1-\left(\frac{tan^{-1}u}{u}\right)^2}{\frac{1+a}{a}\left(\frac{u^4}{5} - \frac{u^2}{3} - \frac{tan^{-1}u}{u} + 1\right)}\right] \tag{5}$$

where ΘD is the Debye temperature, Ω is the molar volume, υ is the velocity of sound, and Γ is a disorder scaling parameter that depends on mass and strain field fluctuations ($\Delta m/m$ and $\Delta\alpha/\alpha$),where α is the ratio of normal-to-Umpklapp prosess [26, 27]. **Figure 1d** shows both the calculated lattice thermal for (PbTe)-(PbSe), (PbTe)-(PbSe)-(PbS) for experimental data and reference data [20]. The experiment data of $(PbTe)_{0.95\text{-}x}$-$(PbSe)_x$-$(PbS)_{0.05}$ follows the trend of calculating data in which lattice thermal conductivity reduces with a decrease of the PbTe matrix. It is clear from **Figure 1d** that solid solution line lies above the lattice thermal in $(PbTe)_{0.95\text{-}x}$-$(PbSe)_x$-$(PbS)_{0.05}$ system by 0.46 $Wm^{-1} K^{-1}$ for x = 0.20, corresponding to 31%. This means that strong phonon scattering in $(PbTe)_{0.95\text{-}x}$-$(PbSe)_x$-$(PbS)_{0.05}$ did not have a significant contribution from point defect scattering. Therefore, the additional 31% difference between theoretical calculation and experimental data arises from other factors, and this issue will be addressed below.

The reduced lattice thermal conductivity observed in $(PbTe)_{0.95\text{-}x}$-$(PbSe)_x$-$(PbS)_{0.05}$ may be attributed to nanostructuring phenomena. The discovery of nanostructuring

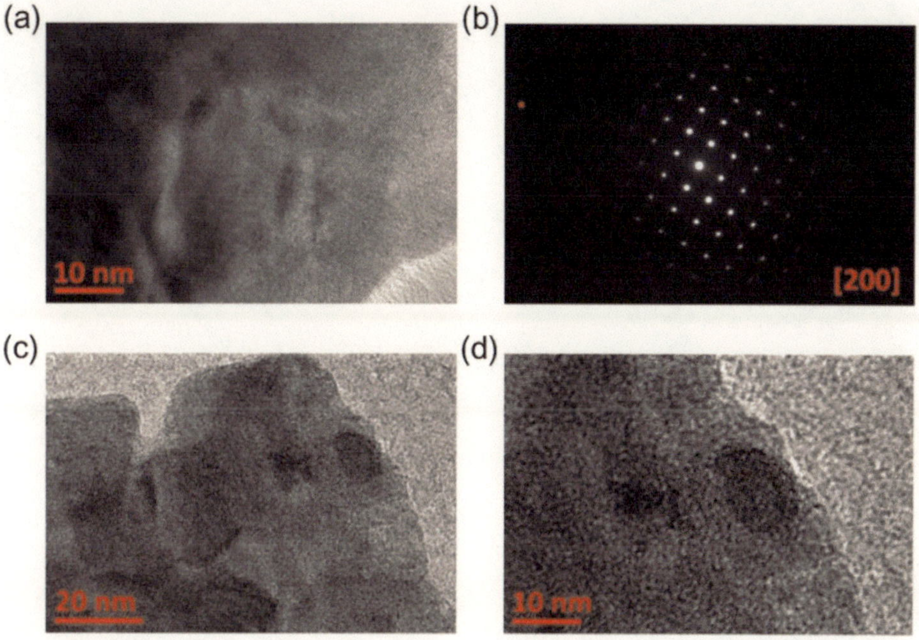

Figure 2.
(a) Low magnification TEM image of $(PbTe)_{0.95}$-$(PbS)_{0.05}$ samples, which reveals nanoscale precipitate with some nucleation and growth. (b) SAED showing single phase material along the [200] zone axis direction. (c) and (d) show low magnification image of $(PbTe)_{0.20}$-$(PbSe)_{0.20}$-$(PbS)_{0.05}$ showing nanostructuring precipitates.

within the $(PbTe)_{0.95\text{-}x}$-$(PbSe)_x$-$(PbS)_{0.05}$ system is unexpected, given its solid solution nature. To explore the plausibility of nanostructuring in lowering the lattice thermal conductivity of $(PbTe)_{0.95\text{-}x}$-$(PbSe)_x$-$(PbS)_{0.05}$, an analysis was performed on $(PbTe)_{0.95}$-$(PbS)_{0.05}$ and samples with the lowest lattice thermal values of $(PbTe)_{0.75}$-$(PbSe)_{0.20}$-$(PbS)_{0.05}$, investigating nanostructuring through TEM characterization.

Figure 2a illustrates a low magnification transmission electron microscopy (TEM) image of $(PbTe)_{0.95}$-$(PbS)_{0.05}$, revealing the presence of small nucleation and growth phenomena. A notable observation is the highly ordered regions suggesting solid solution behavior at a low content of PbS. The 5% PbS content leads to nanostructuring through nucleation and growth mechanisms. This phenomenon is attributed to two primary reasons. First, the nucleating phase must commence with a composite that is not in close proximity to the parent solid solution. Second, the nucleation phase transition exhibits a significant degree of composition change but a minimal extent of size, unlike spinodal decomposition, which demonstrates the opposite characteristics. **Figure 2b** displays a single phase of selected-area diffraction (SAD), indicating that the precipitates share a similar crystal structure and closely matched lattice parameters with the matrix, suggesting an endotoxic relationship. **Figure 2c** and **d** presents typical low magnification TEM images of $(PbTe)_{0.75}$-$(PbS)_{0.20}$-$(PbS)_{0.05}$. The nanostructured precipitates range in size from 5 to 10 nm, showcasing the highest number density of nanoscale precipitates. Irregular morphologies are consistently observed for all nanostructured precipitates, aligning with a nucleation and growth mechanism.

High-resolution transmission electron microscopy (HRTEM) was also utilized for $(PbTe)_{0.75}$-$(PbSe)_{0.20}$-$(PbS)_{0.05}$. Precipitates with a size of 10 nm within the PbTe matrix are depicted in **Figure 3a**. **Figure 3b** and **c** provides middle and high

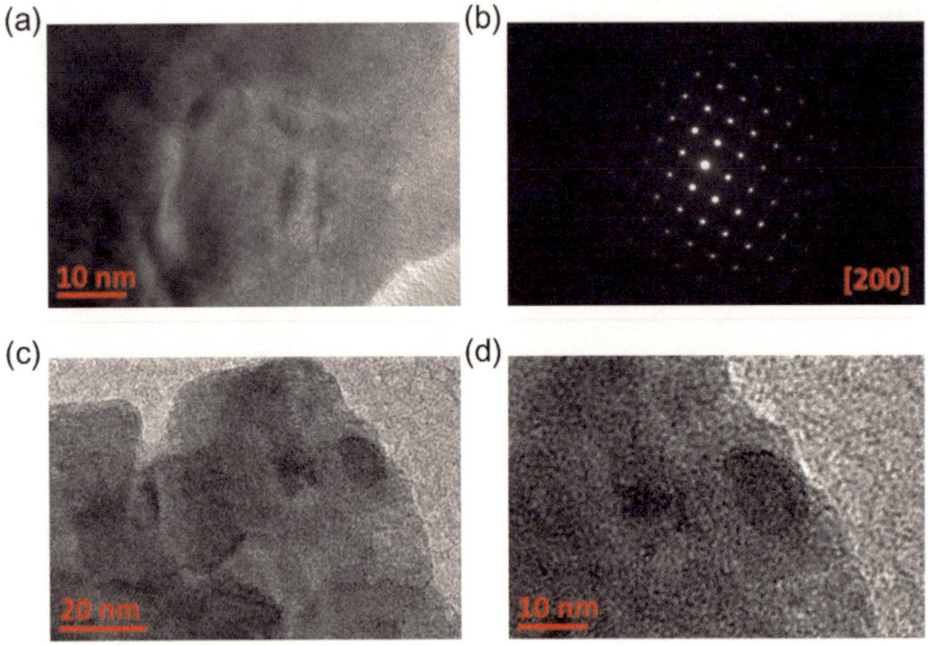

Figure 3.
(a) High resolutions of TEM for $(PbTe)_{0.75}$-$(PbSe)_{0.20}$-$(PbS)_{0.05}$ showing nanoprecipate with matrix. (b) SAED pattern with [200] zone axis direction. (c) and (d) show nanoprecipate of $(PbTe)_{0.75}$-$(PbSe)_{0.20}$-$(PbS)_{0.05}$.

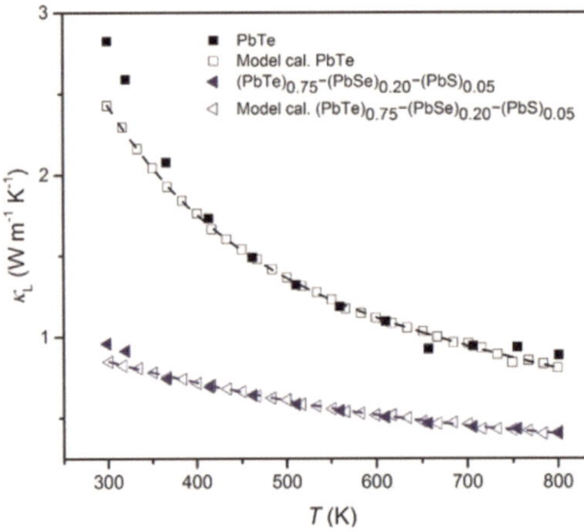

Figure 4.
Theoretical lattice thermal conductivity (lines) of PbTe and $(PbTe)_{0.95}$-$(PbSe)_{0.20}$-$(PbS)_{0.05}$ taking into account that precipice and alloying scattering match considerably with the experimental data (symbols). The data of PbTe and $(PbTe)_{0.95}$-$(PbSe)_{0.20}$-$(PbS)_{0.05}$ was calculated based on the volume fraction PbTe and nanostruturing precipitates with the phenomenological effective medium theory, which described based on Callaway model.

magnification images for $(PbTe)_{0.75}$-$(PbSe)_{0.20}$-$(PbS)_{0.05}$, clearly displaying nanoscale precipitates with irregular morphologies. The volume fraction of precipitates (6.2%) significantly exceeds the nominal PbS presence, indicating that the precipitates do not represent a phase-separated PbS phase. This observation implies that the nanostructuring is a result of a nucleation and growth mechanism.

Previous studies have demonstrated that nanostructuring in the PbTe-PbS system may occur through precipitation, contingent upon its specific compositions and temperature. Within the PbTe-PbS system, the formation of nanostructures can be attributed to either spinodal decomposition or nucleation and growth mechanisms, which are influenced by the system's composition and temperature. This behavior is further elucidated by the presence of a miscibility gap in the PbTe-PbS system, as defined by the Gibbs free energy (G) along a given isotherm within the composition range of x [18, 28–30]. Composition aims to achieve complete homogeneity. In cases where the free energy diagram exhibits positive curvature ($\partial 2G/\partial 2X > 0$), the free energy of the metastable solid solution alloy can be reduced by forming distinct nuclei of a secondary phase, a process known as nucleation and growth. Hence, nucleation and growth phenomena may occur in the solid solution alloy of PbTe-PbS once the energy barrier necessary for these processes is surpassed.

In order to enhance comprehension regarding the correlation between microstructure and lattice thermal conductivity in $(PbTe)_{0.75}$-$(PbSe)_{0.20}$-$(PbS)_{0.05}$, we proceed to offer theoretical computations of the lattice thermal conductivity utilizing Callaway's model [31–33]:

$$\kappa_{lat} = \frac{k_B}{2\pi^2 v}\left(\frac{k_b T}{\hbar}\right)^3 \left\{ \int_0^{\theta_D/T} \tau_c \frac{x^4 e^4}{(e^x - 1)^2} dx + \frac{\left[\int_0^{\theta_D/T} \frac{\tau_c}{\tau_N} \frac{x^4 e^4}{(e^x-1)^2} dx\right]^2}{\int_0^{\theta_D/T} \frac{1}{\tau_N}\left(1 - \frac{\tau_c}{\tau_N}\right) \frac{x^4 e^4}{(e^x-1)^2} dx} \right\} \quad (6)$$

where k_B *is* the Boltzmann's constant, h is the Plank's constant, $x = \omega/_{kBT}$, T is absolute temperature, and ν and θ are speed to sound and Debye temperature, respectively. Here, τ_N is the relaxation time due to normal phonon-phonon scattering, and τ_c is the combined relaxation time using given as:

$$\tau_c^{-1} = \tau_U^{-1} + \tau_N^{-1} + \tau_B^{-1} + \tau_S^{-1} + \tau_D^{-1} + \tau_P^{-1} \quad (7)$$

where τ_U, τ_N, τ_B, τ_S, τ_D, and τ_P are the relaxation time corresponding to scattering from Umklapp process, normal process, boundaries, strain, dislocations, and precipitate [8, 34–38]. Based on these formulae and the parameters obtained from TEM observation (average precipitate size 10 nm, precipitate density 'volume of the nanoparticle'), and some parameters from ref. [36]. We have calculated the theoretical lattice parameter in $(PbTe)_{0.75}$-$(PbSe)_{0.20}$-$(PbS)_{0.05}$. **Figure 4** shows theoretical lattice parameters based on Callaway's. With experimental data. The theoretical data fit well with the experimental data, although the room temperature is somewhat lower than the experimental data. **Figure 4** proves that the strong phonon scattering seems to arise mainly from nanostructuring precipitates compared to other processes due to shorter relaxation time, which means that they play a more dominant role in reducing lattice thermal conductivity in $(PbTe)_{0.75}$-$(PbSe)_{0.20}$-$(PbS)_{0.05}$.

3. Conclusions

This study successfully reveals practical strategies for optimizing low thermal conductivity in PbTe-based materials through nanocomposite and alloy approaches.

The quaternary system $(PbTe)_{0.95-x}$-$(PbSe)_x$-$(PbS)_{0.05}$ with varying PbSe concentrations and a 1 at% Na dopant shows a significant reduction in lattice thermal conductivity. The lowest value was observed in the composition $(PbTe)_{0.75}$-$(PbSe)_{0.20}$-$(PbS)_{0.05}$, achieving a thermal conductivity of 0.42 $Wm^{-1} K^{-1}$, approaching the glass limit for bulk PbTe. Transmission electron microscopy (TEM) analysis identified nanostructuring as a crucial mechanism for the observed reduction in thermal conductivity. The formation of nanostructured precipitates enhances phonon scattering, contributing to lower thermal conductivity.

Acknowledgements

This work was supported by Kyung Hee University (20242252).

Conflicts of interest

There are no conflicts of interest to declare.

Author details

Dianta Ginting[1*] and Jong-Soo Rhyee[2*]

1 Department of Mechanical Engineering, University Mercu Buana, Jakarta Barat, Indonesia

2 Department of Applied Physics and Institute of Natural Sciences, Kyung Hee University, Yongin, South Korea

*Address all correspondence to: dianta.ginting@mercubuana.ac.id
and jsrhyee@khu.ac.kr

References

[1] Mustafa M, Rashak AF, Roy A, Mallick TK, Tahir AA. Advancing thermoelectric materials: A comprehensive review exploring the significance of one-dimensional nano structuring. Nanomaterials. 2023;**13**(13): 2011-2011. DOI: 10.3390/nano13132011

[2] Chung IW. Plainly fixing crystal lattices. Science. 2023;**380**(6647): 800-800. DOI: 10.1126/science.adi2174

[3] Wang B, Zhao H, Wang D, Song A, Chen C, Qin F, et al. Enhanced thermoelectric performance of n-type PbTe via carrier concentration optimization over a broad temperature range. ACS Applied Materials & Interfaces. 2023;**15**:14424-14432. DOI: 10.1021/acsami.3c00465

[4] Ginting D, Lin CC, Lydia R, So HS, Lee H, Hwang J, Kim W, Al Orabi RAR, Jong-Soo. High thermoelectric performance in pseudo quaternary compounds of (PbTe)0.95-x(PbSe)x (PbS)0.05 by simultaneous band convergence and nanoprecipitation. Acta Materialia. 2017;**131**:98-109. DOI: 10.1016/j.actamat.2017.03.036

[5] Snyder GJ. Application of the compatibility factor to the design of segmented and cascaded thermoelectric generators. Applied Physics Letters. 2004;**84**(13):2436-2438. DOI: 10.1063/1.1689396

[6] Wang H, Pei Y, LaLonde AD, Snyder GJ. Heavily doped p-type PbSe with high thermoelectric performance: An alternative for PbTe. Advanced Materials. 2011;**23**:1366-1370. DOI: 10.1002/ADMA.201004200

[7] Wang H, Gibbs ZM, Takagiwa Y, Snyder GJ. Tuning bands of PbSe for better thermoelectric efficiency. Energy and Environmental Science. 2014;**7**(2): 804-811. DOI: 10.1039/C3EE43438A

[8] Wang H, Schechtel E, Pei Y, Snyder GJ. High thermoelectric efficiency of n-type PbS. Advanced Energy Materials. 2013;**3**(4):488-495. DOI: 10.1002/AENM.201200683

[9] Tian Z, Garg J, Esfarjani K, Shiga T, Shiomi J, Chen G. Phonon conduction in PbSe, PbTe, and PbTe1−*x*Se*x* from first-principles calculations. Physical Review B. 2012;**85**(18):184303. DOI: 10.1103/PhysRevB.85.184303

[10] Pei Y, Shi X, LaLonde A, Wang H, Chen L, Snyder GJ. Convergence of electronic bands for high performance bulk thermoelectrics. Nature. 2011;**473**: 66-69. DOI: 10.1038/nature09996

[11] Su T, Li S, Zheng Y, Li H, Hu M, Ma H, et al. Thermoelectric properties of PbTe1−xSex alloys prepared by high pressure. Journal of Physics and Chemistry of Solids. 2013;**74**(7):913-916. DOI: 10.1016/j.jpcs.2013.01.032

[12] Girard SN, He J, Zhou X, Shoemaker D, Jaworski CM, Uher C, et al. High performance Na-doped PbTe–PbS thermoelectric materials: Electronic density of states modification and shape-controlled nanostructures. Journal of the American Chemical Society. 2011; **133**(41):16588-16597. DOI: 10.1021/ja206380h

[13] Wu D, Zhao L-D, Tong X, Li W, Wu L, Tan Q, et al. Superior thermoelectric performance in PbTe-PbS pseudo-binary: Extremely low thermal conductivity and modulated carrier concentration. Energy & Environmental Science. 2015;**8**:2056-2068. DOI: 10.1039/C5EE01147G

[14] Wu HJ, Zhao LD, Zheng FS, Wu D, Pei YL, Tong X, et al. Broad temperature plateau for thermoelectric figure of merit ZT>2 in phase-separated PbTe0.7S0.3. Nature Communications. 2014;**5**:4515. DOI: 10.1038/ncomms5515

[15] Androulakis J, Todorov I, He J, Chung D-Y, Dravid V, Kanatzidis M. Thermoelectrics from abundant chemical elements: High-performance nanostructured PbSe–PbS. Journal of the American Chemical Society. 2011; **133**:10920-10927. DOI: 10.1021/ja203022c

[16] Wang H, Wang J, Cao X, Snyder GJ. Thermoelectric alloys between PbSe and PbS with effective thermal conductivity reduction and high figure of merit. Journal of Materials Chemistry A. 2014; **2**:3169-3174. DOI: 10.1039/c3ta14929c

[17] Pei YL, Liu Y. Electrical and thermal transport properties of Pb-based chalcogenides: PbTe, PbSe, and PbS. Journal of Alloys and Compounds. 2012; **514**:40-44. DOI: 10.1016/j.jallcom.2011.10.036

[18] Girard SN, He J, Li C, Moses S, Wang G, Uher C, et al. In situ nanostructure generation and evolution within a bulk thermoelectric material to reduce lattice thermal conductivity. Nano Letters. 2010;**10**(8):2825-2831. DOI: 10.1021/nl100743q

[19] He J, Girard SN, Kanatzidis MG, Dravid VP. Microstructure-lattice thermal conductivity correlation in nanostructured PbTe0.7S0.3 thermoelectric materials. Advanced Functional Materials. 2010;**20**:764-772. DOI: 10.1002/adfm.2009019

[20] Korkosz RJ, Chasapis TC, Lo S-H, Doak JW, Kim YJ, Wu C-I, et al. High ZT in p-type (PbTe)1–2x(PbSe)x(PbS)x thermoelectric materials. Journal of the American Chemical Society. 2014;**136**:3225-3237. DOI: 10.1021/ja4121583

[21] Aminorroaya Yamini S, Wang H, Ginting D, Mitchell DRG, Dou SX, Snyder GJ. Thermoelectric performance of n-type (PbTe)0.75(PbS)0.15(PbSe)0.1 composites. ACS Applied Materials & Interfaces. 2014;**6**:11476-11483. DOI: 10.1021/am5022257

[22] Aminorroaya Yamini S, Wang H, Gibbs ZM, Pei Y, Mitchell DRG, Dou SX, et al. Thermoelectric performance of tellurium-reduced quaternary p-type lead–chalcogenide composites. Acta Materialia. 2014;**80**:365-372. DOI: 10.1016/j.actamat.2014.06.065

[23] Koh YK, Vineis CJ, Calawa SD, Walsh MP, Cahill DG. Lattice thermal conductivity of nanostructured thermoelectric materials based on PbTe. Applied Physics Letters. 2009;**94**:153101. DOI: 10.1063/1.3117228

[24] Klemens PG. Thermal resistance due to point defects at high temperatures. Physics Review. 1960;**119**:507-509. DOI: 10.1103/PhysRev.119.507

[25] Abeles B. Lattice thermal conductivity of disordered semiconductor alloys at high temperatures. Physics Review. 1963; **131**(5):1906-1911. DOI: 10.1103/PhysRev.131.1906

[26] Adachi S. Lattice thermal resistivity of III–V compound alloys. Journal of Applied Physics. 1983;**54**(4):1844-1848. DOI: 10.1063/1.332820

[27] Adachi S. Lattice thermal conductivity of group-IV and III–V semiconductor alloys. Journal of Applied Physics. 2007;**102**(6):063502. DOI: 10.1063/1.2779259

[28] Darrow MS, White WB, Roy R. Phase relations in the system PbS-PbTe. AIME Metallurgical Society Transactions. 1966. Available from: https://aimehq.org/resources/digital-library

[29] Volykhov AA, Yashina LV, Shtanov VI. Phase relations in pseudobinary systems of germanium, tin, and lead chalcogenides. Inorganic Materials. 2006;**42**:596-604. DOI: 10.1134/S0020168506060045

[30] Darrow MS, White WB, Roy R. Phase relations in the system PbS-PbTe. Materials Science and Engineering. 1969; **3**:289-292

[31] Androulakis J, Lin C-H, Kong H-J, Uher C, Wu C-I, Hogan T, et al. Spinodal decomposition and nucleation and growth as a means to bulk nanostructured thermoelectrics: Enhanced performance in Pb1-xSnxTe−PbS. Journal of the American Chemical Society. 2007;**129**:9780-9788. DOI: 10.1021/ja071875h

[32] Callaway J, von Baeyer HC. Effect of point imperfections on lattice thermal conductivity. Physics Review. 1960;**120**: 1149-1154. DOI: 10.1103/PhysRev.120.1149

[33] Biswas K, He J, Zhang Q, Wang G, Uher C, Dravid VP, et al. Strained endotaxial nanostructures with high thermoelectric figure of merit. Nature Chemistry. 2011;**3**:160-166. DOI: 10.1038/nchem.955

[34] Kim W, Majumdar A. Phonon scattering cross section of polydispersed spherical nanoparticles. Journal of Applied Physics. 2006;**99**:084306. DOI: 10.1063/1.2188251

[35] Kim W, Singer SL, Majumdar A, Zide JMO, Klenov D, Gossard AC, et al. Reducing thermal conductivity of crystalline solids at high temperature using embedded nanostructures. Nano Letters. 2008;**8**:2097-2099. DOI: 10.1021/nl080189t

[36] Mingo N, Hauser D, Kobayashi NP, Plissonnier M, Shakouri A. Nanoparticle-in-alloy approach to efficient thermoelectrics: Silicides in SiGe. Nano Letters. 2009;**9**:711-715. DOI: 10.1021/nl8031982

[37] Minnich AJ, Lee H, Wang XW, Joshi G, Dresselhaus MS, Ren ZF, et al. Modeling study of thermoelectric SiGe nanocomposites. Physical Review B. 2009;**80**:155327. DOI: 10.1103/PhysRevB.80.155327

[38] Prasher R. Thermal transport due to phonons in random nano-particulate media in the multiple and dependent (correlated) elastic scattering regime. Journal of Heat Transfer. 2006;**128**: 627-637. DOI: 10.1115/1.2194036

Chapter 2

Finite Element Formulation of Nonlinear Thermal Conductivity Problems with Applications to Nuclear Fusion

José Miguel Palomino Cobo, Roberto Palma Guerrero, María Esther Puertas García and Rafael Gallego Sevilla

Abstract

Nuclear fusion energy is a highly promising solution to meet future global energy demands. However, its realization requires the development of materials capable of withstanding the extreme conditions in next-generation nuclear fusion reactors. One major challenge is the nonlinear thermal conductivity of these materials. This chapter presents a thermodynamically consistent formulation within the finite element (FE) method for modeling nonlinear thermal conductivity problems. While the formulation is broadly applicable to any nonlinear thermal problem, it is specifically applied in this chapter to materials engineering for nuclear fusion energy, focusing on Eurofer97 material.

Keywords: finite element, heat transfer, nonlinear thermal conductivity, nuclear fusion, computational mechanics

1. Introduction

Accurate analysis of thermal problems is crucial for the design and optimization of components subjected to temperature gradients. Furthermore, in most engineering applications, thermal conductivity exhibits material nonlinearity, namely, a dependence on temperature. Consequently, this nonlinearity is a challenge for the computational mechanics community and requires advanced methods to capture the true thermal response of materials.

For simple geometries, boundary conditions, and temperature-dependent material properties, analytical or semi-analytical solutions can be derived using techniques such as Boltzmann and Kirchhoff transformations, Taylor series expansions, Laplace transforms, or perturbative methods [1]. For more complex problems, numerical methods, such as boundary element, finite difference, and finite element (FE) methods, are required.

This chapter addresses these challenges using the finite element (FE) method, with a focus on temperature-dependent thermal properties, particularly thermal conductivity, which plays a crucial role in heat transfer and temperature distribution. The FE method has been extensively applied to complex thermal problems, including nonlinear heat conduction in multiphysics scenarios, such as thermoelectric materials, photothermal conversion in solar nanofluids, and thermal stress analysis in phase change materials [2–6]. This chapter extends the previous FE formulations by incorporating heterogeneous heat sources for nuclear heating in fusion energy applications.

The remainder of this chapter is organized as follows: Section 2 introduces nuclear fusion materials, with a focus on their temperature-dependent thermal properties. Section 3 presents the strong and weak forms of the nonlinear heat equation, along with the FE discretization. Finally, Section 4 reports four validation examples: (1) a comparison between analytical and numerical results for two-dimensional, linear, and steady-state problem; (2) a numerical analysis of one-dimensional and nonlinear material behavior; (3) a study of transient nonlinear thermal responses; and (4) a real-world application involving nuclear fusion materials. These examples demonstrate the effectiveness of the proposed method for capturing complex thermal behaviors in engineering applications.

2. Nuclear fusion materials

In the context of the DONES program, which aims to test and qualify materials for the construction of future fusion power reactors, the IFMIF-DONES project [7] is a DEMO reactor designed for material testing, and it will be constructed in Granada, Spain.

Reduced activation ferritic/martensitic (RAFM) steels are regarded as the most advanced and ready-to-use materials for fusion reactor construction due to their proven manufacturing and welding capabilities, as well as the extensive industrial experience gained in their application [8]. These materials must meet stringent requirements, such as high radiation resistance and thermal stability. Among them, Eurofer97 is a notable class of steel alloys specifically engineered to address the challenges posed by nuclear fusion. The term Eurofer, derived from “European Fusion Experimental Reactor,” refers to a family of steels meticulously designed to endure the extreme conditions within fusion reactors.

Although the details of the IFMIF-DONES project are beyond the scope of this chapter, it is important to note that several designs for the irradiated capsules are being proposed [9]. This type of irradiation generates heterogeneous nuclear heating within the capsules, which must be considered. Additionally, one of the materials inside the capsules, the Eurofer97 [10], exhibit strong temperature dependence [11]:

Density [kg/m^3]:

$$\rho = 7750 \tag{1}$$

Thermal conductivity [W/mK]:

$$\kappa(T) = 0.190706\,T - 4.3053 \cdot 10^{-4} T^2 + 3.817 \cdot 10^{-7} T^3 - 1.158 \cdot 10^{-10} T^4 \tag{2}$$

Thermal capacity [J/kgK]:

$$C(T) = 2.696\,T - 0.00496\,T^2 + 3.303 \cdot 10^{-6} T^3 \tag{3}$$

Within the IFMIF-DONES project, temperature variations in components involving Eurofer97 can reach approximately 400 K. Consequently, the use of sophisticated nonlinear models is essential, as simplifying these effects with a linear model could lead to significant inaccuracies.

3. Nonlinear heat equation

The heat equation is used to analyze heat transfer and temperature changes over time. In linear problems, variable relationships are proportional, but in nonlinear problems, they become more complex because of temperature-dependent material properties, boundary conditions (such as radiation), or phase changes [1].

As discussed in Section 2, the thermal properties of materials have a polynomial relationship with temperature, requiring numerical techniques such as the FE method [12] for an approximate solution. This technique begins with the weak form of the heat equation, where the solution is found in a space of test functions, selecting the one that satisfies the boundary conditions with minimal error [13]. This allows for flexible approximations and high accuracy by discretizing the domain into finite elements [14].

This section addresses the nonlinear heat conduction problem in three parts (or subsections): strong forms, weak forms, and FE discretization.

3.1 Strong form

The governing equations commonly comprise balance equations, constitutive relations, and boundary conditions. For the thermal problem, these are as follows:

The heat transfer equation, derived from the principle of conservation of energy, reads

$$\rho c(T)\dot{T} = -q_{i,i} + Q(x_i, t), \tag{4}$$

where Q denotes the heterogeneous heat source.

The transport equation (commonly called Fourier's law), obtained from the entropy balance, becomes:

$$q_i = -\kappa(T)T_{,i}. \tag{5}$$

Essential or Dirichlet type boundary conditions: $T = \overline{T}$, where $\overline{T}$ denotes the prescribed temperature.

Natural or Newmann type boundary conditions, including prescribed thermal flux $\overline{q}$ and convection and radiation phenomena:

$$q_i n_i = \overline{q} + h(T - T_0) + \eta\sigma(T - T_0)^4, \tag{6}$$

where n denotes the outward normal to the boundary, h is the convection coefficient, η the emissivity, σ the Stefan-Boltzmann constant, and T_0 is the external temperature. The radiative heat flux is nonlinear, whereas the convective flux is linear. However, in the following of the present chapter, the radiation and convection terms are not considered, as the primary focus is to illustrate the application of the numerical method.

For the sake of clarity, this chapter uses the index notation summarized in the Appendix A. Additionally, the material properties relevant to the thermal problem are as follows:

- Density ρ represents mass per unit volume.
- Thermal conductivity κ, indicating that the material can transfer heat.
- Thermal capacity c, expressing the thermal energy required to raise the temperature of a unit mass by one degree.

3.2 Weak form

A classical solution in partial differential equations (PDEs) is smooth and continuous in the domain. Classical solutions are often easy to find and provide an accurate description of the phenomenon modeled by the PDE. However, some PDEs are difficult to solve due to boundary or initial conditions or may not have classical solutions because of singularities in the domain. In these cases, the weak or variational formulation of PDE is used as an alternative to find a solution that satisfies conditions in an integral sense, rather than at every point.

In general, the weak formulation of any problem is derived as follows [15]:

1. Multiply the balance equation by a test or variation function. For the heat problem, the test variation is denoted as δT, and it refers to a small perturbation in the temperature distribution of the domain.
2. Integrate over the domain of interest Ω with boundary $\Gamma = \partial\Omega$.
3. Apply the Green's theorem.
4. Introduce the boundary conditions into the equation.

Applying the above steps, the functional to minimize becomes

$$\int_{\Omega} \delta T q_{i,i} \mathrm{d}\Omega + \int_{\Omega} \delta T \left(\rho c \dot{T} - Q\right) \mathrm{d}\Omega = 0. \tag{7}$$

The first term is further developed using Green's theorem and the chain rule as follows:

$$\int_{\partial\Omega} \delta T q_i \cdot n_i d\Gamma = \int_{\Omega} \left(\delta T q_i\right)_{,i} d\Omega = \int_{\Omega} \delta T_{,i} q_i d\Omega + \int_{\Omega} \delta T q_{i,i} d\Omega, \tag{8}$$

where an expression for the first term of eq. (7) is obtained. In the present chapter, the surface integral is not considered for the sake of simplicity and (7) becomes

$$-\int_{\Omega} \delta T_{,i} q_i d\Omega + \int_{\Omega} \delta T \left(\rho c \dot{T} - Q\right) d\Omega = 0 \tag{9}$$

The above equation serves as the basis for the numerical discretization. Note that the problem was presented using the weighted residuals method; however, it could also be obtained using a variational method. Both methods yield similar results but differ in mathematical approaches. While variational theorems are elegant for minimizing terms such as energy, the weighted residuals method is more general and flexible, making it suitable for a wider range of problems.

3.3 Finite element discretization

As commented, the FE method is a powerful numerical technique used to approximate solutions of PDEs. Generally, finite elements are used to discretize the domain, while finite differences are used for time discretization.

For the thermal problem, the domain of interest Ω must first be approximated by a computational domain Ω_h and discretized into N_{el} finite elements of domain Ω_e:

$$\Omega \simeq \Omega_h = \bigcup_{e}^{N_{el}} \Omega_e. \tag{10}$$

Each *finite element* is a smaller and simpler region for which the solution $T(x_i, t)$ is approximated using shape functions $N_i(x_i)$ as follows:

$$T(x_i, t) = N_a(\xi_i)\overline{T}^a(t) \tag{11}$$

where $q = 1, \ldots, n$ denotes the number of nodes within each element and $\overline{T}^a(t)$ are the nodal values of the approximation, which are unknown and depend on time t. The interpolation functions $N_a(\xi_i)$ depends on the natural coordinates $\xi_i \in [-1, 1]$ of each element and must satisfy the following condition:

$$\sum_{a=1}^{n} N_a(\xi_i) = 1. \tag{12}$$

The derivatives can be constructed using the chain rule and expressed as

$$\frac{\partial N_a}{\partial \xi_i} = \frac{\partial N_a}{\partial x_j}\frac{\partial x_j}{\partial \xi_i} = \frac{\partial N_a}{\partial x_j} J_{ji}, \tag{13}$$

where J_{ji} denotes the Jacobian of the transformation between the natural and global coordinates. Using the derivatives of the shape functions, the temperature gradient can be discretized as

$$T_{,i} = N_{a,i}\overline{T}^a. \tag{14}$$

For simplicity, a Galerkin approach is used, where variations are approximated in the same way as the solution:

$$\delta T = N_a(\xi_i)\delta\overline{T}^a \Rightarrow \delta T_{,i} = N_{a,i}\delta\overline{T}^a. \tag{15}$$

Introducing these approximations in the weak form of (9), the residual function at each node a of the element with domain Ω_e is expressed as follows:

$$R_a\left(\overline{T}^a, \dot{\overline{T}}^a\right) = -\int_{\Omega_e} N_{a,i} q_i d\Omega_e + \int_{\Omega_e} N_a\left(\rho c \dot{\overline{T}}^a - Q\right) d\Omega_e. \tag{16}$$

By considering two different nodes a, b of a particular element, the residual can be expressed as a function of the tangent stiffness matrix K_{ab} tangent capacity matrix C_{ab} as follows:

$$R_a\left(\overline{T}^a, \dot{\overline{T}}^a\right) = -\left(K_{ab}\overline{T}^b + C_{ab}\dot{\overline{T}}^b\right) + F_a, \tag{17}$$

where F_a denotes the external sources. For clarity, the iteration numbers for the Newmark-β algorithms "k" are omitted below. The tangent matrices can be obtained from (17) as

where F_a denotes the external sources. The temporal evolution of the system is addressed using the Newmark-β method, where the residual is linearized and minimized at each time step n iteratively using the following tangent matrix:

$$S_{ab^{(k)}} = \alpha_1 K_{ab^{(k)}} + \alpha_2 C_{ab^{(k)}} \tag{18}$$

Here, the parameters α_1 and α_2 are scalar coefficients defined by the Newmark scheme, which depend on the discretization parameters and ensure the stability and accuracy of the method.

The solution process at time t_n starts using the temperature field $\overline{T}^a$ and its time derivatives from the previous time step t_{n-1} as an initial guess. The temperature is then updated using the Newmark-β method. With the updated temperature, the tangent matrix is evaluated, and the residual is estimated based on the linearization. If the residual norm does not decrease below a specified tolerance ε (e.g., $|R| < \varepsilon$), the process proceeds iteratively with the updated temperature until convergence is achieved. For further information, see [17].

$$K_{ab} = -\frac{\partial R_a}{\partial \overline{T}^b}, \qquad C_{ab} = -\frac{\partial R_a}{\partial \dot{\overline{T}}^b}. \tag{19}$$

As commented, the material properties depend on temperature, which results in material nonlinearity. In the first approximation, the properties exhibit a polynomial dependence on temperature within the temperature range where the simulation occurs:

$$\kappa(T) = \kappa_0 + \kappa_1 T + \kappa_2 T^2, \qquad c(T) = c_0 + c_1 T + c_2 T^2 + c_3 T^3, \tag{20}$$

On the one hand, the tangent stiffness matrix can be obtained from the residual of (16) as follows:

$$K_{ab} = -\frac{\partial R_a}{\partial \overline{T}^b} = \int_{\Omega_e} N_{a,i} \frac{\partial q_i}{\partial \overline{T}^b} \partial\Omega_e - \int_{\Omega_e} N_a \rho \frac{\partial c(T)}{\partial \overline{T}^b} \partial\Omega_e, \tag{21}$$

where the derivatives can be explicitly calculated as follows:

$$\frac{\partial q_i}{\partial \overline{T}^b} = -\kappa(T)\frac{\partial T_{,i}}{\partial \overline{T}_b} - \frac{\partial \kappa(T)}{\partial T}\frac{\partial T}{\partial \overline{T}^b} T_{,i} = -\kappa(T) N_{b,i} - \frac{\partial \kappa(T)}{\partial T} N_b T_{,i}. \tag{22}$$

$$\frac{\partial c(T)}{\partial \overline{T}^b} = \frac{\partial c(T)}{\partial T} N_b, \tag{23}$$

and introducing into eq. (21):

$$K_{ab} = -\int_{\Omega_e} N_{a,i}\left(\frac{\partial\kappa}{\partial T}N_b T_{,i} + \kappa(T)N_{b,i}\right)\partial\Omega_e - \int_{\Omega_e} N_a\rho\left(\frac{\partial c}{\partial T}N_b\right)\partial\Omega_e. \tag{24}$$

On the other hand, the tangent capacity matrix is obtained from the residual of Eq. (16) to give

$$C_{ab} = -\frac{\partial R_a}{\partial \dot{T}^b} = -\int_{\Omega_e} N_a\rho c(T)N_b\partial\Omega_e. \tag{25}$$

From Eqs. (24) and (25), the tangent matrix can be obtained by equation (19).

According to Taylor and Govindjee [16], the numerical integration is performed by Gauss-Legendre quadrature [17]:

$$\iiint_{-1}^{+1} f(x_1, x_2, x_3)\,\mathrm{d}x_1\,\mathrm{d}x_2\,\mathrm{d}x_3 \simeq \sum_{l=1}^{L} f(\zeta_l)W_l, \tag{26}$$

where L is the number of ζ_l integration (also called Gauss) points and W_l is the weight associated with each integration point. For example, the capacity matrix can be integrated as follows [15]:

$$C_{ab} = \iiint_{-1}^{+1} N_a(\bar{\xi}_i)\rho c(\overline{T})N_b(\bar{\xi}_i)J\ \mathrm{d}\bar{\xi}_i \simeq \sum_{l=1}^{L} N_a(\bar{\xi}_l)\rho c(T)N_b(\bar{\xi}_l)JW_l. \tag{27}$$

Here, the key point is that all elements are parametrized identically. Thus, repeated node positions in the mesh have the same natural coordinates, and the integrals share weights and integration points.

Finally, the nonlinear and dynamic formulation is implemented in a user subroutine of the FEAP software [18], which belongs to the University of California at Berkeley (USA). This software offers the following advantages:

1. Nonlinearities are solved using advanced Newton-Raphson or Line Search algorithms.

2. Time integration is performed using backward finite differences.

3. The global tangent matrices are assembled by the software.

4. The boundary conditions for fluxes and temperatures are prescribed by specific FEAP commands.

In summary, the user is only required to implement the formulation at the element level. As outlined in the next section, the code was thoroughly tested using various examples to ensure robustness and applicability. The code is currently being applied to analyses related to the IFMIF-DONES project.

4. Results

This section presents four examples of the numerical accuracy of the nonlinear FE method discussed in this chapter. In particular, the examples are:

- Example 1: two-dimensional linear steady-state problem.
- Example 2: one-dimensional nonlinear steady-state problem.
- Example 3: two-dimensional nonlinear transient problem.
- Example 4: application to Eurofer97.

4.1 Example 1: Validation of linear heat conduction, analytical vs. numerical solutions

This example compares the analytical and numerical solutions for a two-dimensional domain without heat sources under the assumption of linear thermal conductivity and steady-state conditions.

The domain of the problem is defined as a rectangle in the (x, y)-plane, with dimensions 1×1 [m]. Only Dirichlet boundary conditions are considered: $T = \overline{T} = 1$ [K] is prescribed on the upper boundary $y = 1$ and $T = \overline{T} = 0$ at the remaining three boundaries ($x = 0$, $x = 1$, and $y = 0$).

For this example, the governing equations are reduced to the Laplace equation $\nabla^2 T(x,y) = 0$. An analytical solution can be derived using the separation of variables and Fourier series as follows:

$$T^{analytical}(x,y) = \frac{4\overline{T}}{\pi}\sum_{k=1}^{\infty}\frac{\sin\left(\frac{(2k-1)\pi x}{l_2}\right)\sinh\left(\frac{(2k-1)\pi y}{l_2}\right)}{(2k-1)\sinh\left(\frac{(2k-1)\pi l_3}{l_2}\right)}. \tag{28}$$

This analytical solution provides a benchmark for the FE solution, thereby allowing the accuracy of the FE implementation to be quantified by comparing both the numerical and closed solutions. **Figure 1** shows the contour plots of the analytical (left) and numerical (right) solutions. The contour plot closely resembled the analytical solution, confirming the accuracy of the code for linear thermal conduction.

The accuracy of the FE solution is quantified by comparing both solutions by the L_2-norm of the error defined as:

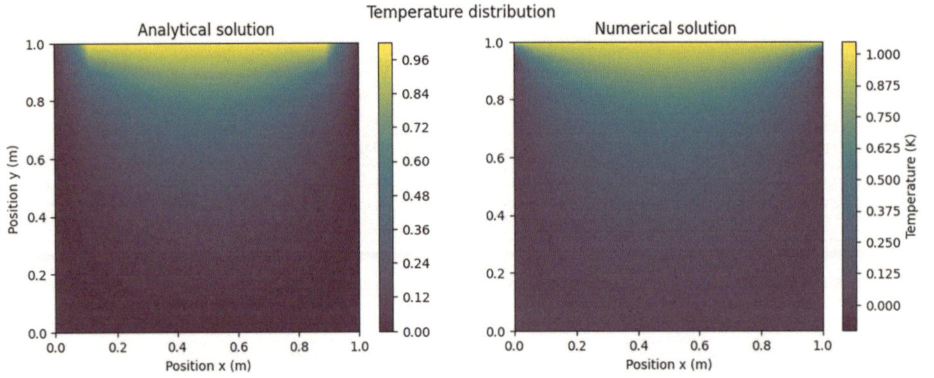

Figure 1.
Contour plots for analytical (left) and numerical (right) solutions.

Mesh size	N	E_{L_2} error	Remarks
5x5	25	0.093	Initial coarser mesh
10x10	100	0.045	Coarse mesh
20x20	400	0.022	Improved resolution
50x50	2500	0.008	High-resolution mesh

Table 1.
Error analysis between analytical and FE solutions for various mesh sizes.

$$E_{L_2} = \sqrt{\sum_{a=1}^{N} \left(T_a^{\text{fem}} - T_a^{\text{theo}}\right)^2}, \tag{29}$$

where N denotes the total number of nodes in the mesh. As reported in **Table 1**, the L_2-norm of the error decreases with increasing mesh size, demonstrating convergence of the FE solution toward the analytical one.

4.2 Example 2: Non-linear heat conduction, dimensional approximation and numerical validation

To validate the accuracy of the FE code for solving nonlinear problems with heat sources, a one-dimensional analytical solution was compared against the FE results. In particular, the analytical solution considers a thin slab of thickness L subject to Q, where the thermal conductivity depends on temperature in the polynomial form:

$$\kappa(T) = k_0\left(1 + \beta T + \gamma T^2\right). \tag{30}$$

Given the temperature at the mid-plane of the slab, denoted as T_0, the closed solution is obtained through a Taylor series expansion under the assumptions of steady-state, one-dimensional conduction, mid-plane symmetry, and uniform energy generation, as outlined in [1]. Mathematically, the distribution of temperature reads

$$T(x) = T_0 - \frac{Qx^2}{2k_0\left(1 + \beta T_0 + \gamma T_0^2\right)} - \frac{3Q^2(\beta + 2\gamma T_0)}{k_0^2\left(1 + \beta T_0 + \gamma T_0^2\right)^3} \frac{x^4}{4!}. \tag{31}$$

Then, for the linear case $k = k_0$, the analytical solution is reduced as follows:

$$T(x) = T_0 - \frac{Qx^2}{2k_0}. \tag{32}$$

Numerically, a two-dimensional domain of 0.01×0.025 [m] is considered, centered on the x-axis, and with 10 four-noded elements along this axis. Adiabatic flux conditions are applied at the upper and lower boundaries. For this example, $T_0 = 323.0\,\text{K}$ and $Q = 100\ \text{W/m}^3$, while the coefficients of the nonlinear conductivity are as follows:

$$\beta = 3.6 \cdot 10^{-3}, \quad \gamma = 2.5 \cdot 10^{-6}, \quad k_0 = 2 \cdot 10^{-4}\,\text{W/mK}. \tag{33}$$

The FE results for the linear and nonlinear behaviors are shown in **Figure 2** (left and right, respectively).

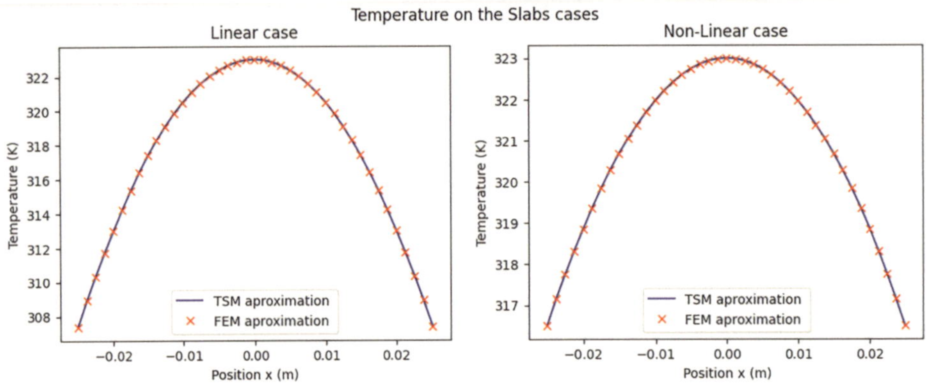

Figure 2.
Temperature distribution on the slab for the linear (left) and nonlinear (right) cases.

Despite the initial assumption that the nonlinear terms may be insignificant because of their near-zero values before solving the problem, **Figure 2** shows that this assumption is incorrect. The comparison revealed a significant difference in the temperature distribution between the two cases. For instance, in the linear case, there is a temperature difference of nearly 20 degrees between the center and edges of the slab, whereas in the nonlinear case, the difference is only around 6 degrees.

Again, the global L^2 error [19] is used to quantify the precision of the approximation. For the linear case, the error value was approximately 7.45×10^{-5}, while for the nonlinear, it becomes 3.7×10^{-4}. As expected, the nonlinear case exhibits a higher error due to the complexity of modeling temperature-dependent thermal conductivity.

4.3 Example 3: Dynamic non-linear heat transfer

The main objective of this example is to evaluate the accuracy of the nonlinear and dynamic code by comparing the FE solutions with those obtained using the KDC-GFDM approach in [20]. Specifically, this example considers a square domain of size 1×1 [m] with temperature-dependent thermal conductivity, prescribed heat flux at the boundary, and an internal heat source.

Firstly, the thermal properties for this problem are $\kappa(T) = 0.3 \cdot T + 150\,\mathrm{W/m^{\circ}C}$, $Q = 1000\ \mathrm{W/m^3}, \rho = 7850\,\mathrm{kg/m^3}$, and $c = 460\,\mathrm{J/kg^{\circ}C}$. Secondly, the boundary and initial conditions are, respectively, specified as follows:

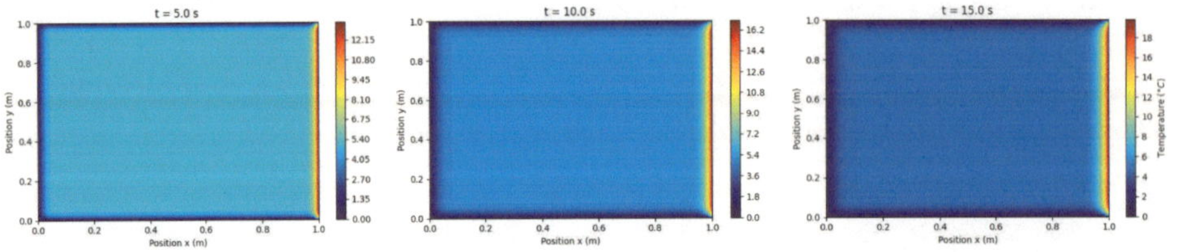

Figure 3.
Results generated by our FEM code: temperature distributions in the square domain at t = 5, 10, and 15 s.

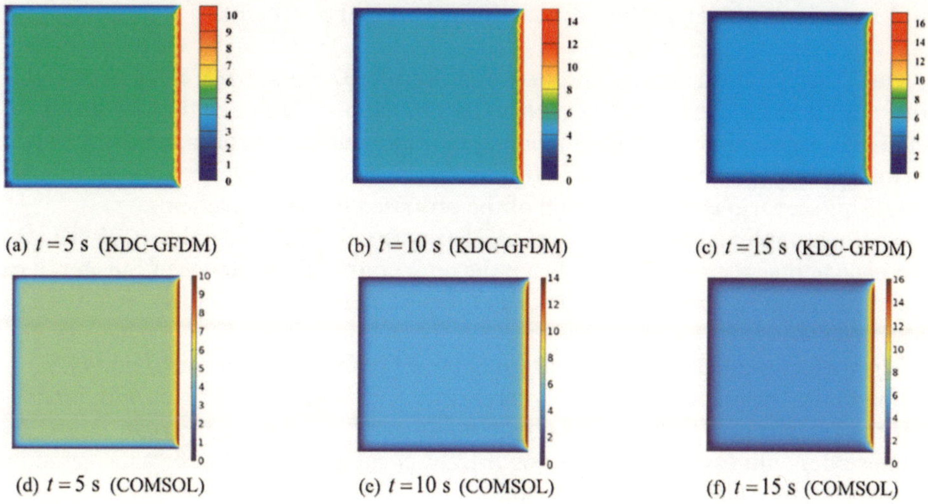

Figure 4.
Comparison of temperature distributions in the square domain at t = 5, 10, and 15 s. Results computed by Comsol and KDC-GFD Method and taken from [20].

- A heat flux $q = 80{,}000\,\mathrm{W/m^2}$ on the right edge.
- A temperature $T = 0^{\circ}\mathrm{C}$ on all other edges.
- The initial condition across the domain is $T(x, y, 0) = 5^{\circ}\mathrm{C}$.

Numerical solutions are computed over the interval $t \in [0, 15]$ [s], with time discretized using a step size of $\Delta t = 0.1$ [s]. The results for $t = 5{,}10{,}15$ [s] are shown in **Figure 3**. After comparing these results with those reported in [20] shown in **Figure 4**, it is concluded that the present FE code shows good agreement for this nonlinear and transient problem.

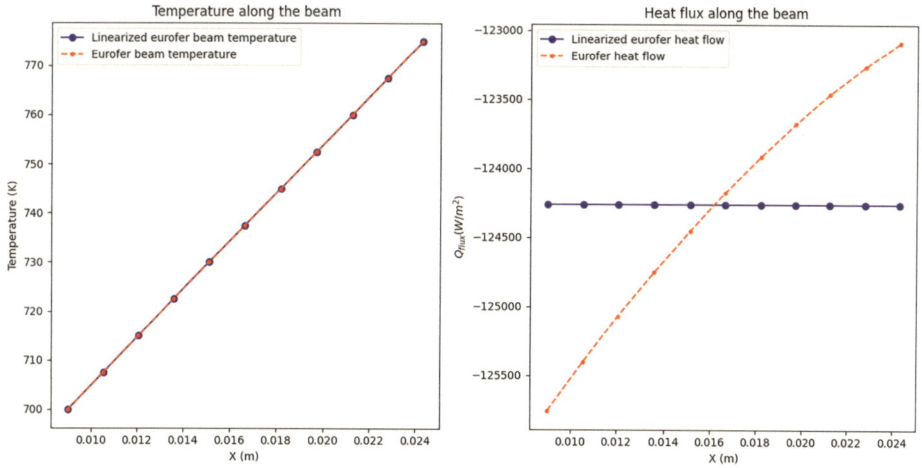

Figure 5.
Temperature (left) and heat flux (right) for the two models of Eurofer97.

4.4 Example 4: Linearization of non-linear thermal problems

Once the nonlinear thermal properties of Eurofer97 are known, a key question arises: can they be linearized to simplify the simulations? Linearization would be beneficial, especially for studies involving multiple Eurofer97 components, such as reliability analysis and design optimization, because it would reduce computational costs. This scenario is an application of the proposed numerical scheme.

These examples simulate an Eurofer97 bar with cold 700 [K] and hot 775 [K] and a reference temperature of 725 [K]. From the temperature dependence of Eurofer, it can be assumed that its thermal conductivity remains constant with a mean value of $\overline{\kappa} =$ 25.35 [W/mK].

Figure 5 presents the numerical results for $\overline{\kappa}$ (denoted as linearized) and $\kappa(T)$. Although the temperature distribution remains practically identical for both models (left figure), the heat flux exhibited high nonlinearity, as seen in the right figure. In conclusion, nonlinearity introduces significant changes that should be considered in engineering and scientific applications.

5. Conclusions

Many problems in science and engineering exhibit nonlinear thermal behavior due to the temperature dependence of material properties. Ignoring these nonlinearities can lead to inaccurate conclusions, especially in applications with significant temperature variations.

Although many commercial software packages can handle certain nonlinear thermal problems, it is essential to explore coding alternatives to reduce the reliance on such tools in future projects. In addition, advanced applications, such as uncertainty quantification, sensitivity analysis, reliability studies, and Bayesian inverse problems [21], can often be tackled more effectively using custom codes than commercial software.

In this context, this chapter addressed the complexity of nonlinear thermal problems and highlighted the importance of temperature-dependent thermal properties. A robust numerical methodology based on the finite element method is presented and demonstrated through four illustrative examples.

Acknowledgements

This research was funded by the European Commission under the EUROFUSION2 Project, within the framework of the Solid Mechanics and Structures research line – LI/EUROFUSION-Gallego, and supported by Grant Agreement No. 101052200.

The views and interpretations expressed here are the responsibility of the authors and do not necessarily represent those of the European Commission.

Conflict of interest

The authors declare no conflict of interest.

Appendix A. Index notation in tensor theory

The index notation represents equations and mathematical expressions using indices rather than traditional algebraic symbols. This notation is commonly used in tensor theory, continuum mechanics, general relativity, and other areas of physics and mathematics where it is necessary to handle vector and tensor quantities.

In index notation, indices indicate the components of vectors, tensors, or matrices on a specific basis. The indices can take numerical values, such as 1, 2, 3, to represent components in Cartesian coordinates (x, y, z), or they can be letters, such as i, j, k, to represent components on a general basis. The basic rules of index notation are as follows:

1. *Einstein summation convention*: When an index is repeated in an expression, a summation of all possible index values is implied. For example, consider the expression $a_i b_i$; the Einstein summation convention implies that it actually represents $a_1 b_1 + a_2 b_2 + a_3 b_3$ in three-dimensional space.

2. *Index repetition convention*: When an index appears in both the upper and lower parts of a derivative, such as in $\frac{\partial x_i}{\partial x_j}$, summation over the repeated index is implied. In this example, it represents $\frac{\partial x_1}{\partial x_j} + \frac{\partial x_2}{\partial x_j} + \frac{\partial x_3}{\partial x_j}$ in three-dimensional space.

3. *Index contraction:* When an index appears in the same position in two terms, summation over that index is implied. For example, in $T_{ij} x_j$, it is represented as $T_{i1} x_1 + T_{i2} x_2 + T_{i3} x_3$ in three-dimensional space.

Index notation is especially useful when working with higher order tensors because it allows equations to be expressed more compactly and generally, regardless of the coordinates used. It is important to note that index notation can be somewhat more complex to read initially, but with practice and understanding of the basic rules, it becomes a powerful tool for the mathematical manipulation of tensor and vector expressions.

Author details

José Miguel Palomino Cobo*, Roberto Palma Guerrero, María Esther Puertas García and Rafael Gallego Sevilla
Department of Structural Mechanics and Hydraulic Engineering, University of Granada (UGR), Granada, Spain

*Address all correspondence to: jmpalominocobo@ugr.es

References

[1] Jiji LM. Non-Linear Conduction Problems. Berlin, Heidelberg: Springer Berlin Heidelberg; 2009. pp. 215-235. DOI: 10.1007/978-3-642-01267-9_7

[2] Palma R, Pérez-Aparicio JL, Taylor RL. Non-linear finite element formulation applied to thermoelectric materials under hyperbolic heat conduction model. Computer Methods in Applied Mechanics and Engineering. 2012;**213-216**:93-103. Available from: https://www.sciencedirect.com/science/article/pii/S0045782511003574

[3] Palma R, Moliner E, Pérez-Aparicio JL. Elasto-thermoelectric beam formulation for modeling thermoelectric devices. Finite Elements in Analysis and Design. 2017;**129**:32-41

[4] Forner-Escrig J, Mondragon R, Hernandez L, Palma R. Non-linear finite element modelling of light-to-heat energy conversion applied to solar nanofluids. International Journal of Mechanical Sciences. 2020;**188**:105952

[5] Forner-Escrig J, Palma R, Mondragón R. Finite element formulation to study thermal stresses in nanoencapsulated phase change materials for energy storage. Journal of Thermal Stresses. 2020;**43**(5):543-562

[6] Palma R, Moliner E, Forner-Escrig J. Computational Thermoelectricity Applied to Cooling Devices. In: Aranguren P, editor. Bringing Thermoelectricity into Reality. Vol. 13. Rijeka: IntechOpen; 2018

[7] IFMIF/EVEDA. High Flux Test Module. 2016. Available from: https://www.ifmif.org/high-flux-test-module/

[8] Baluc N, Gelles D, Jitsukawa S, Kimura A, Klueh R, Odette G, et al. Status of reduced activation ferritic/martensitic steel development. Journal of Nuclear Materials. 2007;**367**:33-41

[9] Álvarez I, Anguiano M, Mota F, Hernández R, Qiu Y. Neutronic assessment of the IFMIF-DONES HFTM specimen stack distribution. Fusion Engineering and Design. 2024;**200**:114212. Available from: https://www.sciencedirect.com/science/article/pii/S0920379624000656

[10] Gordeev S, Schwab F, Arbeiter F, Qiu Y. Numerical study of conjugated heat transfer for DONES high flux test module. Fusion Engineering and Design. 2019;**09**:146

[11] Mergia K, Boukos N. Structural, thermal, electrical and magnetic properties of Eurofer 97 steel. Journal of Nuclear Materials. 2008;**373**(1):1-8. Available from: https://www.sciencedirect.com/science/article/pii/S0022311507006642

[12] Zienkiewicz OC, Taylor RL, Zhu JZ. The Finite Element Method: Its Basis and Fundamentals. Oxford, UK: Butterworth-Heinemann; 2005

[13] Lax PD. Hyperbolic Partial Differential Equations. Vol. 14. Providence, Rhode Island, USA: American Mathematical Society; 2006

[14] Mercier B. An Introduction to the Numerical Analysis of Spectral Methods. New York, NY, USA: Springer; 1989

[15] Taylor R, Govindjee S. FEAP—A finite element analysis program, Theory Manual. Berkeley, California, USA: University of California at Berkeley, USA; 2020. Available from: http://projects.ce.berkeley.edu/feap/theory86.pdf

[16] Taylor R, Govindjee S. FEAP—A finite element analysis program, Programmer Manual. Berkeley, California 94720-1710: University of California at Berkeley, USA; 2020. Available from: http://projects.ce.berkeley.edu/feap/pmanual 86.pdf

[17] Press WH. Numerical recipes 3rd edition: The art of scientific computing. Cambridge, UK: Cambridge University Press; 2007

[18] Taylor R, Govindjee S. FEAP—A finite element analysis program, User Manual. Berkeley, California, USA: University of California at Berkeley, USA; 2020. Available from: http://projects.ce.berkeley.edu/feap/manual 86.pdf

[19] Hesthaven JS, Warburton T. Nodal discontinuous Galerkin methods: algorithms, analysis, and applications. New York, NY, USA: Springer Science and Business Media; 2007

[20] Sun W, Ma H, Qu W. A hybrid numerical method for non-linear transient heat conduction problems with temperature-dependent thermal conductivity. Applied Mathematics Letters. 2024;**148**:108868

[21] Betz W, Papaioannou I, Straub D. Bayesian post-processing of Monte Carlo simulation in reliability analysis. Reliability Engineering System Safety. 2022;**227**:108731. Available from: https://www.sciencedirect.com/science/article/pii/S0951832022003544

Chapter 3

Heat Transfer Enhancement on Staggered Perforated Circular Pin-Fin Heat Sink: An Experimental Assessment

Ndah Abdulrahman Alpha and Aondoyila Kuhe

Abstract

This experimental study examines how forced convective flow affects heat transfer properties in a rectangular channel with staggered pin fins featuring different perforation patterns under constant heat flux conditions across Reynolds numbers (Re) ranging from 2.0×10^3 to 12×10^3. The study compares cylindrical pin fins with solid pin fins and those featuring circular longitudinal (L), longitudinal/transverse (LT), and longitudinal/transverse/vertical (LTV) perforations to determine optimal perforation configurations for enhanced heat transfer performance. The experiment uses a Peltier module to generate heat on one side, utilizing the Armfield Free and Forced Convection Heat Transfer Service Units HT 19 and HT10XC. The results showed that perforated pins significantly raise Nusselt number (Nu) over solid pins: 7% for L, 30% for LT, and 64% for LTV perforations. Pressure drops are reduced by 10% for L, 17% for LT, and 25% for LTV perforations relative to solid pins. At lower Reynolds numbers, the overall enhancement ratio peaks, notable for reaching a 40% rise with LTV-perforated pin fins. Additionally, fin effectiveness improves significantly: 14, 34, and 57% higher for L, LT, and LTV perforated pin-fin arrays, respectively. This study showcases potential applications in electronic cooling systems, promising improved heat transfer efficiency.

Keywords: pin-fin heat sink, Peltier module, perforation pattern, heat transfer coefficient, electronic cooling

1. Introduction

The relentless push for higher performance in microelectronics, coupled with the drive for miniaturization, has made efficient thermal management a paramount concern in modern electronic design. As electronic devices become more compact and powerful, the ability to dissipate heat effectively is crucial to ensuring their performance, reliability, and longevity. The relationship between component temperature and the operational efficacy of electronic devices is well established–lower temperatures generally equate to better performance and a longer lifespan for these components.

Heat sinks play a critical role in managing the thermal loads of electronic devices. By facilitating heat transfer from heat-generating components, heat sinks help maintain temperatures within safe operating limits. The challenge lies in designing heat sinks that not only efficiently dissipate heat but also fit within the increasingly constrained spaces of modern electronics.

Perforated fins are especially effective at improving heat transfer because of their greater ratio of surface area to volume, resulting in lighter heat sinks and more uniform temperature distribution compared to solid fins. Research consistently shows that heat sinks with perforated pin-fins outperform those with solid pin-fins in terms of thermal efficiency [1–5]. Studies on various pin-fin configurations further underscore these findings [6–10].

Additional studies have further underscored the advantages of specific perforation geometry. Ibrahim et al. [11] highlighted the benefits of triangular perforations in reducing pressure drop and increasing the Nusselt number, while Chingulpitak et al. [12] noted that enlarging both the quantity and size of perforations in pin-fins results in better thermal performance.

Chin et al. [1] demonstrated that perforations enhance convection heat transmission by 45% and reduce pressure drop by 18% compared to solid pins. Moreover, perforations reduce the heat sink's weight by 37%, improving overall heat transfer. Kore et al. [13] found that conical perforations in pin-fin heat sinks significantly enhance flow turbulence and heat transmission. According to Gupta et al. [14], circular perforations improve thermohydraulic performance by 28%, with a temperature reduction of approximately 16 K when using square-shaped perforations.

Other studies have investigated the effects of varying perforation sizes and quantities on the efficiency of heat sinks. Choure et al. [15] found that enlarging both the quantity and diameter of perforations leads to a reduced pressure drop and an enhancement in the Nusselt number. Foo et al. [16] found that perforated pin-fin arrays, subjected to varying inlet velocities, transfer more heat and experience a lower pressure drop compared to solid pin arrays.

Hayder et al. [17] examined the influence of pin geometry on thermal efficiency, concluding that round pin fins facilitate better heat transfer than those with sharp edges. Similarly, Kotcioglu et al. [18] discovered that circular pin fins demonstrate a lower pressure drop in comparison with hexagonal or square pin fins. Hwang et al. [19] proposed a thermal optimization model demonstrating that varying fin thickness can reduce thermal resistance, particularly in water-cooled heat sinks. Koga et al. [20] designed an enhanced heat sink using the Topology Optimization Method tailored for small-scale applications, with an emphasis on reducing pressure drop while maximizing thermal dissipation.

In the existing literature on heat sink optimization through perforations, most studies have focused on individual types of perforations [8, 12, 21–24], resulting in a fragmented understanding of the effectiveness of circular perforated cylindrical fin pin heat sinks.

Building on this extensive foundation, the current investigation seeks to push the boundaries of heat sink design by examining the impact of various perforation configurations on heat transfer efficiency. This research aims to identify the optimal perforation pattern that maximizes heat dissipation efficiency. Experimental analysis will be conducted on heat sinks with solid pin-fins, longitudinal perforations, longitudinal/transverse perforations, and longitudinal/transverse/vertical perforations across different inlet velocities and Reynolds numbers. Key performance metrics, including the Nusselt number, friction factor, thermal resistance, hydrothermal performance, fin efficiency, and effectiveness, are evaluated to determine the most effective design.

In essence, this study aspires to advance the state of the art in thermal management for electronic devices, providing insights that could lead to more efficient and compact cooling solutions. By systematically exploring and analyzing the effects of various perforation configurations, this research aims to contribute valuable knowledge to the ongoing quest for optimal heat sink designs.

2. Methods

The experiment was conducted using the Armfield Unit for Heat Transfer by Free and Forced Convection, designed to demonstrate the principles of natural (free) and forced convection. As depicted in **Figure 1**, this unit features a bench-mounted vertical rectangular air duct (300 × 350 mm and 950 mm in height) positioned above a centrifugal fan. Inside the duct, a hot wire anemometer measures and displays the air

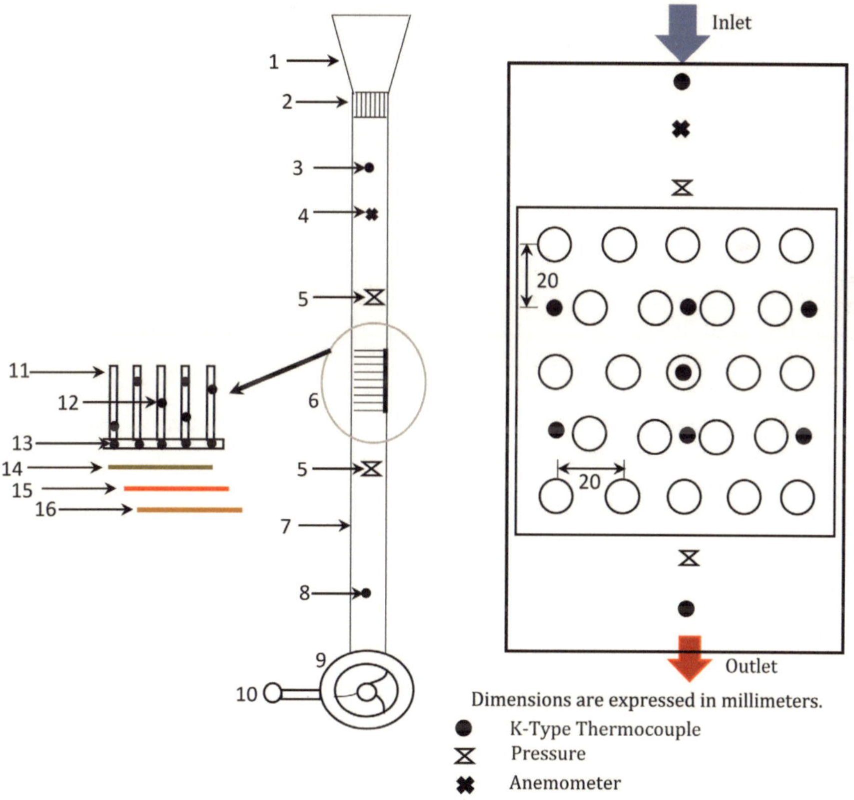

Figure 1.
Diagram illustrating the experimental setup. – Components include the mouth (1), anti-turbulent screen (2), inlet thermocouple (3), anemometer (4), inlet Pitot tube (5), test section (6), air duct (7), outlet thermocouple (8), variable speed centrifugal fan (9), throttle (10), test specimen (heat sink) (11), surface thermocouples (12), base plate thermocouples (13), thermal interface material (14), thermoelectric device (15), and Teflon insulation (16).

velocity on a dedicated display unit. The duct includes a 150 × 150 mm test section at the rear wall, allowing the insertion of various heat-transfer surfaces.

Each heat sink unit is equipped with a polycarbonate sheet with low thermal conductivity at its base, serving as an insulator to ensure upward heat flow through the heat sink. A 60-watt Peltier module (heating pad) with stabilized heat flux mimics the thermal heating of electronic components. The heat sink, characterized by high thermal conductivity, is connected to the heating pad using high-conductivity thermal tape to minimize contact resistance. Thermocouples positioned on the heat sink's base measure the temperature across the surface.

The duct's front wall is constructed from acrylic to allow clear observation of the heated surface and measurement sensors. A throttle that can be adjusted manually regulates the airflow, while pressure sensors placed both upstream and downstream of the orifice plate monitor the difference in pressure across the test section. Thermocouples measure air temperatures before and after the heated surface, as well as base and surface temperatures at various points along the extended surface of the heat exchangers.

The base plate has seven primary thermocouples (T3–T9) to measure its surface temperature, while the pins have eight secondary thermocouples (T1–T8) to measure extended surface temperatures. Air inlet and outlet temperatures are recorded by sensors T1 and T2 at the duct's top and base. Thermocouple attachment points are covered with adhesive for protection. Air velocity can be adjusted from 0 to 12 m/s, depending on local mains voltage and supply frequency, with the sensor permanently mounted in the duct and connected to the console via a plug and socket. All thermocouples are "Duplex," with primary thermocouples using miniature plugs and secondary thermocouples using two-way edge connectors suitable for the Risepro Data Logger.

Four different configurations of staggered pin arrays have been used, namely:

i. Solid pin-fin heat sink depicted in **Figure 2a**,

ii. Heat sink with longitudinal perforated pin-fins shown in **Figure 2b**,

iii. Heat sink featuring both longitudinal and transverse perforated pin-fins illustrated in **Figure 2c**, and

iv. Heat sink incorporating longitudinal, transverse, and vertical perforated pin-fins as shown in **Figure 2d**.

Each pin-fin array contains 23 pins, 8 mm in diameter and 45 mm in height, with a pitch of 20 mm.

2.1 Experimental procedure

The experiment entailed testing four types of heat sinks at five flow speeds (0.5, 1.0, 1.5, 2.0, and 2.5 meters/second) with a constant heat flux of 5000 Wm^{-2}, resulting in 20 experimental runs using a consistent approach.

2.1.1 Preparation

i. The laboratory air conditioning system was activated to stabilize the room temperature at 25°C.

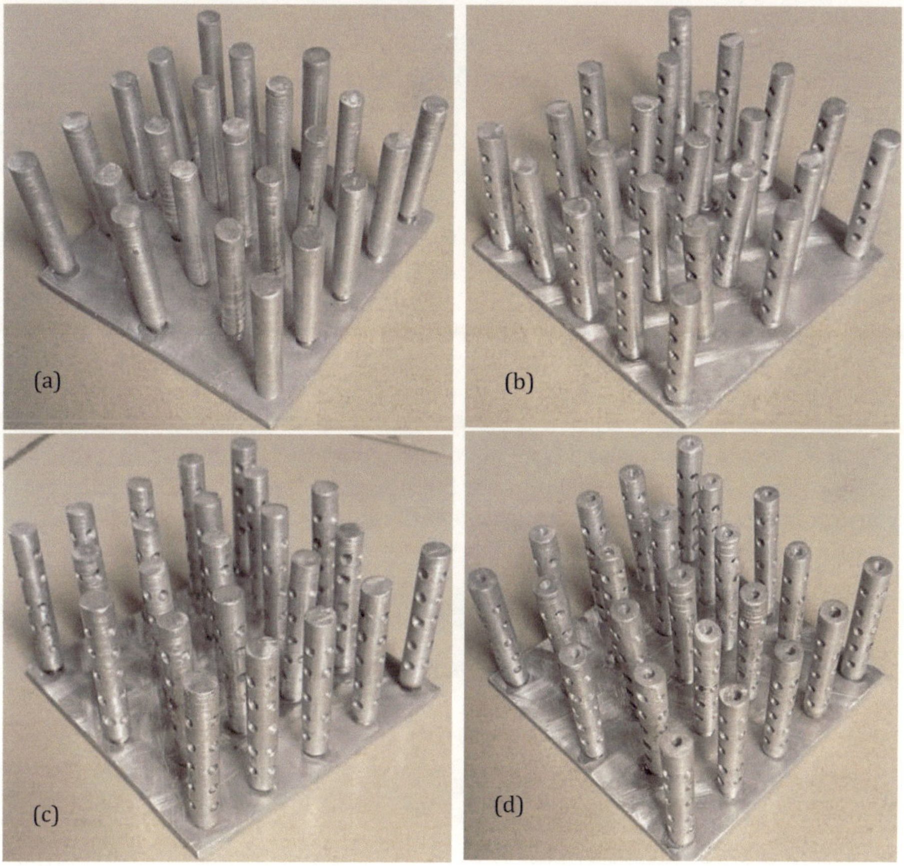

Figure 2.
Different configurations of staggered pin arrays have been used.

ii. The HT10XC heat transfer service equipment was powered on.

iii. The centrifugal ventilator was started to induce fluid flow through the test section. The air velocity at the entrance was adjusted to the desired value and maintained constant.

2.1.2 Measurements

i. The digital differential manometer was zeroed to record the pressure difference.

ii. The variable transformer was set to provide the required input power to the heating pad, monitored via a digital wattmeter to ensure the desired heat flux.

iii. The system was allowed to reach steady-state conditions over approximately 80 minutes, as determined by stable specimen temperatures (fluctuations within ±0.1°C over 10 minutes).

2.1.3 Data collection

i. Air velocity was checked every 10 minutes for consistency.

ii. Thermocouple readings were taken consistently until stabilization, at which point final measurements were recorded.

In steady-state operation, according to the fundamental rule of thermodynamics, energy input must equal energy output [25]. The heat generated by the electrically heated surface ($Q_{electrical}$) is dissipated via convection ($Q_{convection}$), conduction ($Q_{conduction}$), and radiation ($Q_{radiation}$).

$$Q_{electrical} = Q_{convection} + Q_{conduction} + Q_{radiation} \tag{1}$$

Considering that the fins and base plate were composed of aluminum with polished surfaces and operational temperatures kept within reasonable limits, radiative heat loss was negligible, constituting less than 0.5% of the total input [26]. Therefore, the equation simplifies to:

$$Q_{electrical} = Q_{convection} \tag{2}$$

The following is an expression for convective heat transfer:

$$Q_{convection} = h_{av} A_s (T_{s,\mathrm{av}} - T_b), \tag{3}$$

$$T_b = \frac{T_{in} + T_{out}}{2}, \tag{4}$$

Consequently, one can determine the average convective heat transfer coefficient from Eqs. (1)–(4) as follows:

$$h_{av} = \frac{Q_{convection}}{A_s \left[T_{s,av} - \left(\frac{T_{in}+T_{out}}{2}\right)\right]} \tag{5}$$

Where,
T_b = Average bulk temperature.
T_{out} = the exit airflow temperature.
T_{in} = the inlet airflow temperature.
$T_{s,\mathrm{av}}$ = the mean surface temperature of a pin-fin array.
A_s = the pin-fin array's total surface area.
The dimensionless Nusselt (Nu) and Reynolds (Re) numbers are calculated as [25, 27]:

$$Nu = \frac{h_{av} D_h}{k_a} \tag{6}$$

$$Re = \frac{\rho_a V D_h}{\mu_a} = \frac{V D_h}{v_a} \tag{7}$$

Where,
D_h = the channel's hydraulic diameter.
k_a = air's thermal conductivity.
ρ_a = density of air.

μ_a = dynamic air viscosity.
v_a = kinematic viscosity of air.
The rectangular section of the passage's hydraulic diameter is described as. [1, 26, 27].

$$D_h = 4\frac{A_c}{P_c} = \frac{2BW}{(B+W)}, \tag{8}$$

Where,
A_c = tunnel's cross-sectional area.
P = the tunnel's perimeter.
B = Duct cross-section depth.
W =Duct cross-section width.

The friction factor (f) can be determined by measuring the pressure at both ends of the test section using the equation provided in Ref. [25].

$$f = \frac{2\Delta P D_h}{b\rho {V_\infty}^2} \tag{9}$$

Where, ΔP = The decrease in air pressure as it flows across the array of test pin fins.

V = The average velocity of the input channel across its cross-sectional area.

The key parameter determining the thermal efficiency of the pin-fin heat sink is the total thermal resistance, and it is evaluated as follows [28]:

$$R_{th} = \frac{T_{base} + T_\infty}{Q_{convection}}, \tag{10}$$

Considering the same pumping force, the overall improvement factor (hydrothermal performance) is defined as the ratio between the enhancement ratios of thermal transfer and friction coefficients. It is defined by the following expressions [29–32]:

$$\text{Overall enhancement ratio } \eta = \frac{\frac{Nu}{N_s}}{\left(\frac{f}{f_s}\right)^{\frac{1}{3}}}, \tag{11}$$

Where,
Nu = Nusselt number for a heat sink with perforated pin fins.
f = friction factor for a heat sink with perforated pin fins.
Nu_s = Nusselt number for a heat sink with solid pin fins.
f_s = friction factor for a heat sink with solid pin fins.

The fin effectiveness (performance) ε_f is defined as the ratio of the heat transfer by the fin to the heat transfer at the fin base in a constant test section base temperature scenario [25, 33].

It is defined as follows:

$$\varepsilon_f = \frac{h_{av} A_s (T_{s,av} - T_\infty)}{h_{av,base} A_{s,base} (T_{s,av\,base} - T_\infty)} \tag{12}$$

Where,
T_∞ = the free stream temperature.
$h_{av,base}$ = base flat plate average convection heat transfer coefficient.

$A_{s,base}$ = surface area of the base plate.

$T_{s,av\ base}$ =the base average surface temperature of a flat plate.

Another measure of fin thermal performance is fin efficiency η_f, which refers to the ratio of the actual heat transfer rate from the fin to the ideal heat transfer rate that would occur if the entire fin were at the base temperature [25, 33].

It is defined as follows:

$$\eta_f = \frac{h_{av}A_s(T_{s,av} - T_b)}{h_{av}A_s(T_{s,av\ base} - T_b)} \tag{13}$$

3. Results and discussion

The present study investigates the use of novel perforated pin heat sinks with multiple perforations to cool electronic components. Four experimental heat sinks are examined: solid, L, LT, and LTV perforated pin-fin.

3.1 Heat transfer characteristics

The primary objective of designing a perforated heat sink is to maximize heat transfer rates while minimizing energy consumption. **Figure 3** illustrates the correlation between the Nusselt number and the Reynolds number for various pin-fin arrays: solid, L, LT, and LTV perforations. The results demonstrate that the Nu for all heat sink types increases approximately linearly with the Reynolds number. Perforations significantly impact the average heat transfer rate, with transverse and vertical perforations providing substantial improvements. Notably, the Nusselt number for pins with longitudinal (L) perforation is approximately 7% greater than that for solid pins; longitudinal/transverse (LT) perforation is approximately 30% higher; and longitudinal/transverse/vertical (LTV) perforation is around 64% higher.

The staggered arrangement of pins also affects the average heat transfer rate by disturbing the thermal boundary layer and enhancing airflow mixing, which increases the contact area for heat transfer. Among the perforated pin-fins, LTV perforations exhibit the highest increase in the Nusselt number compared to solid pins.

In solid-pin heat sinks, airflow separation creates dead flow zones downstream, leading to higher temperatures and lower Nusselt numbers. Longitudinal perforations

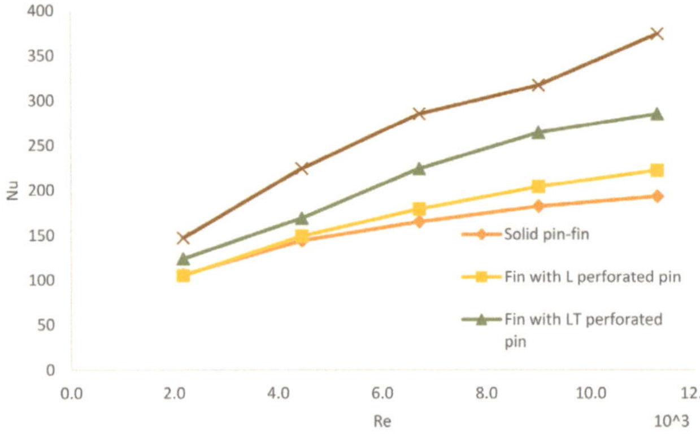

Figure 3.
Nusselt number variation with Re for solid pin fin, L perforations, LT perforations, and LTV perforations.

help mitigate these thermal dead zones by reducing the intensity of the hot zones located behind the pins. These perforations facilitate fluid mixing, delaying flow separation from the pin surface.

Transverse perforations further enhance heat transfer by creating small disturbances in the airflow, which mix high-speed air closer to the wall and maintain attachment of the flow to the surface further downstream, resulting in narrower wakes than those of solid pins. Increased heat transfer resulting from perforations results from both the greater surface area and enhanced local effects near the perforations, facilitated by the formation of localized air jets, consistent with the findings of Ndah et al. [34].

As the number of perforations rises, the mixing of fluid within the pin heat sink also increases, leading to lower pin temperatures due to multi-jet airflow, which in turn raises the Nusselt number. The LTV perforations offer the most significant enhancement for the solid pin fin array. This is because vertical perforations allow additional airflow to pass through, effectively blending with the main airflow to generate more turbulence and extending the contact time between the working fluid and hot surfaces, thereby enhancing the rate of heat transfer. Moreover, the decreased weight of the heat sink resulting from its perforations plays a significant role in reducing processor temperatures, which is crucial for managing thermal conditions in electronics.

3.2 Friction factor (f)

The impact of heat sink designs on airflow behavior is illustrated in **Figure 4**, which clearly demonstrates that the friction factor declines as the Reynolds number rises. Additionally, the friction factor is higher for ducts with solid pin fins compared to those with perforated pin fins. Similar findings have been reported by [22, 34, 35].

The pressure drops across perforated pins is less than that across solid pins. This difference is attributed to the removal of sections from the frontal area of the pins, enabling airflow to pass through the perforations [34, 36]. The solid pins restrict the airflow path, leading to increased friction drag and additional areas of recirculation and separation of airflow downstream. In contrast, perforated designs create a wider and straighter airflow path, resulting in less separation and a lower pressure drop due to the pins' permeability. This aligns with the findings of Ndah et al. [34].

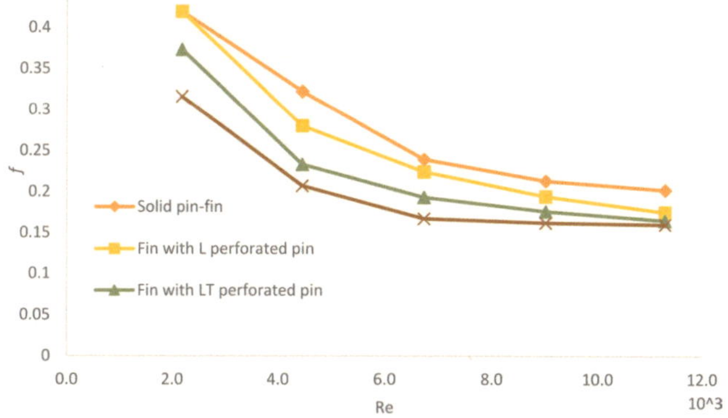

Figure 4.
Friction factor variation with Re for solid pin fin, L perforations, LT perforations, and LTV perforations.

Recirculation regions located downstream of perforated pin fins are smaller than those behind solid pin-fins owing to the transverse perforations. These eddies are more pronounced downstream of solid pins but are significantly reduced. In situations involving perforated pins, airflow passes through the longitudinal and transverse perforations, which delays flow separation downstream, reducing the magnitude of the vortices and, consequently, the pressure drag. Increasing the number of perforations reduces the size of recirculation zones, boosting the efficiency of perforated pin-fin heat sinks. In solid pin fin heat sinks, airflow separation occurs at the pins' frontal surface area.

At a given Reynolds number, an increase in the number of perforations results in a corresponding decrease in pressure drop. For Reynolds numbers greater than 3.0×10^3, pin fins with L, LT, and LTV perforations showed pressure drop reductions of 10, 17, and 25%, respectively, compared to solid pins. This is because solid pins cause more blockage than perforated pins, leading to higher pressure drops. Thus, it is evident that perforated pins result in lower pressure drops.

As shown in **Figure 4**, for each heat sink type (Solid, L, LT, and LTV), as the number of perforations increases, the pressure drop decreases as more air passes through the perforations. The results indicate that LTV-perforated pin fins are particularly effective, requiring less fan power because of their lower friction factor.

3.3 Total thermal resistance (R_{th})

Effective thermal management in electronics, particularly in CPU applications, necessitates maintaining temperatures below the critical threshold of approximately 85°C [37, 38] while minimizing energy costs. A primary objective is to demonstrate the advantages and reliability of perforated pin-fin heat sink designs for typical CPU thermal management.

Optimal heat sink performance is characterized by low thermal resistance and pressure drop values. **Figure 5** illustrates the variation in overall thermal resistance as a function of Re for solid, L, LT, and LTV perforated pin-fin arrays. At a higher Reynolds number (greater than 6000), the thermal resistance for all types of perforated pin-fin arrays experiences a minimal and approximately constant value of less than 0.5°C/W. For the L type, it is around 0.38°C/W, representing a 16.7% reduction compared to the solid type. It is about 0.27°C/W for the LT type, indicating a 33.3%

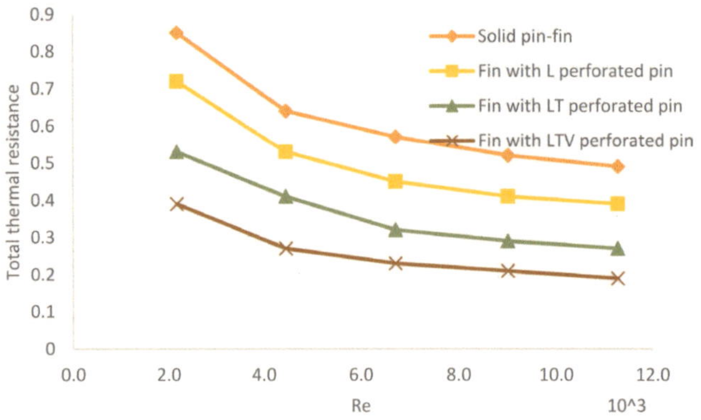

Figure 5.
Total thermal resistance variation with Re for solid pin fin, L perforations, LT perforations, and LTV perforations.

reduction. For the LTV type, it is roughly 0.20°C/W, indicating a 44.4% reduction in thermal resistance. The figure confirms that perforated pins significantly enhance heat transfer, thereby reducing base plate temperatures and thermal resistance. Among the heat sinks examined, the LTV pin-fin heat sink showed superior performance, demonstrating the least thermal resistance. This is attributed to the vertical perforations, which induce turbulence within the fins, enhancing the mixing of buoyant flow. This increased interaction time between the working fluid and hot surfaces enhances the heat transfer coefficient, reduces the base temperature, and consequently lowers thermal resistance. Similar observations were made in related studies [39, 40].

3.4 Overall enhancement ratio (η)

Efficient heat transfer necessitates maximizing heat transfer while minimizing pressure drop across various thermal component configurations. Therefore, an improved design should take into account the balance between heat transfer performance and the resulting pressure drops. This study examines the overall enhancement ratio, which can be calculated for different configuration scenarios across various Re. The overall enhancement ratio measures fin effectiveness under specific boundary conditions.

It has been observed that the overall enhancement ratio for L, LT, and LTV perforated fins exceeds that of solid fins. This finding aligns with the conclusions of Ismail et al. [41]. **Figure 6** demonstrates that the greatest enhancement was achieved with the LTV-perforated pin-fin due to the lower pressure drop from the flow. It is commonly understood that heat transfer enhancement is attained by applying force to drive the flow through the channel, measured as pumping pressure. However, it is crucial to remember that an increase in pumping force does not necessarily result in a proportional gain in the Nusselt number. This is due to the limited time for the flow to effectively carry heat away from the pin-fin array, which explains the decreasing performance factor with an increasing Reynolds number. Therefore, the LTV perforated pin-fin achieved greater heat transfer enhancement (Nu/Nus) and less friction factor compared to solid pins (f/fs), making it the preferred model for pin fins compared to solid, L, and LT perforated pin-fin arrays.

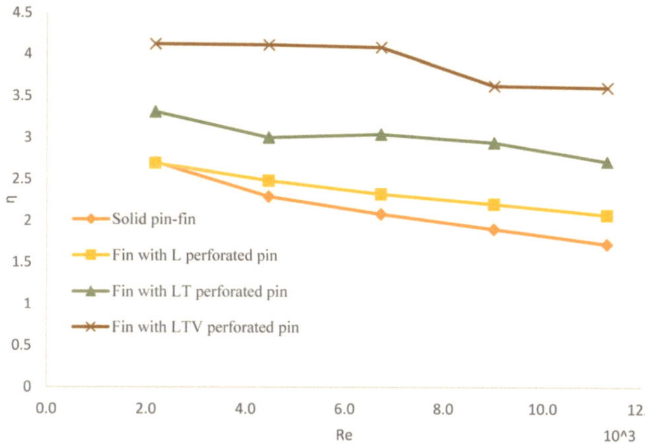

Figure 6.
Effect of solid pin fin, L perforations, LT perforations, and LTV perforations on overall enhancement ratio.

3.5 Fin effectiveness (ε_f)

For a constant base section heat flux, **Figure 7** illustrates the correlation between fin effectiveness and Re for solid, L, LT, and LTV perforated pin-fin arrays. It is evident that the effectiveness of these pin-fin arrays is independent of the Re; however, effectiveness increases with the number of perforations. This observation aligns with Noaman et al.'s [42] findings on heat transfer enhancement through perforated fins and [43, 44].

The data also clearly show that a higher number of perforations results in greater effectiveness, exceeding 4.0, indicating enhanced heat transfer. Additionally, the fin effectiveness for L, LT, and LTV perforated pin-fin arrays is 14, 34, and 57% higher, respectively, than that of the solid pin fin array. This improvement is attributed to the perforations, which reduce recirculation zones behind the solid fins.

3.6 Fin efficiency (η_f)

Figure 8 illustrates the effect of fin efficiency on Re for solid, L, LT, and LTV perforated pin-fin arrays under a constant base section heat flux. The results demonstrate a decline in fin efficiency with increasing Reynolds numbers across all pin-fin arrays, consistent with [42]. For Reynolds numbers less than 2.0×10^3, the fin efficiency of all heat sink types exceeds 0.85%. In particular, the solid pin-fin array's fin efficiency is approximately 0.92. The L-type array shows a fin efficiency of about 0.89, representing a 3.3% decrease compared to the solid type. The LT-type array has an efficiency of approximately 0.85, indicating a 7.6% decrease, whereas the LTV-type array has an efficiency of roughly 0.84, indicating an 8.7% decrease.

The higher fin efficiency of solid pin-fin arrays compared to L, LT, and LTV perforated pin-fin arrays can be attributed to the reduced thermal conduction along the length of the perforated pin fins, which diminishes their efficiency relative to solid pin fins.

In thermal management applications, the perforated pin-fin heat sink designs, particularly the LTV type, offer significant benefits over conventional solid pin-fin

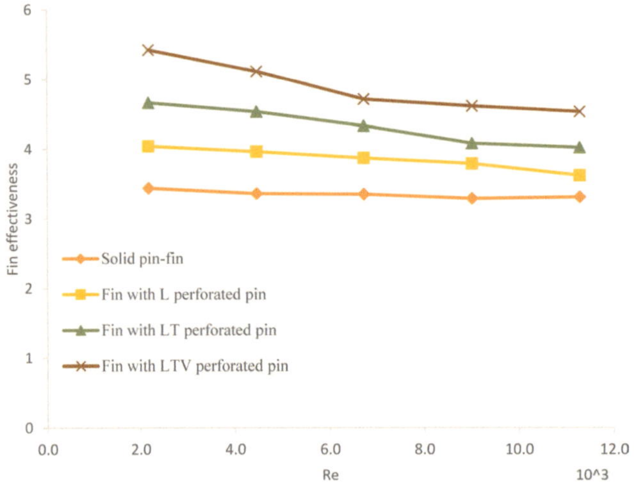

Figure 7.
Effect of solid pin fin, L perforations, LT perforations, and LTV perforations on fin effectiveness.

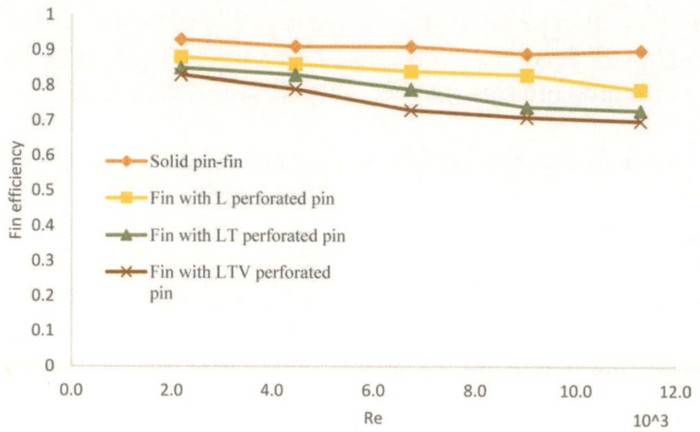

Figure 8.
Effect of solid pin fin, L perforations, LT perforations, and LTV perforations on fin efficiency.

designs. The strategic incorporation of perforations enhances heat transfer efficiency, reduces thermal resistance, and increases the overall effectiveness of the heat sinks, making them highly suitable for applications requiring efficient cooling solutions. The slight reduction in fin efficiency is a trade-off for the substantial gains in thermal performance, positioning perforated pin-fin heat sinks as superior options for advanced thermal management.

4. Conclusions

Experimental analysis is used in this study to investigate several key parameters, including total heat transfer, friction factor, thermal resistance, overall enhancement ratio, fin effectiveness, and fin efficiency. The effects of various configurations, such as solid pin-fin, longitudinal, transverse, and vertical perforated pin-fin arrays, on these parameters are examined. The conclusions drawn from this investigation are summarized as follows:

1. The Nu for longitudinal (L) perforated pins is approximately 9% greater than that of solid pins, while for longitudinal/transverse (LT) perforated pins, it is about 33% higher, and for longitudinal/transverse/vertical (LTV) perforated pins, it rises by approximately 67% in comparison to solid pins.

2. Pressure drop is reduced by approximately 9% for longitudinal (L) perforated pins compared to solid pins, by 19% for longitudinal/transverse (LT) perforated pins, and by 27% for longitudinal/transverse/vertical (LTV) perforated pins relative to solid pins.

3. The hydrothermal performance factor shows the greatest increase, particularly evident (38%) in the case of the LTV perforated pin-fin array at lower Reynolds numbers.

4. The thermal resistance of all types of pin-fin arrays is minimal and approximately constant, with a value of less than 0.5°C/W.

5. The weight reduction percentages relative to solid pin fins are 19% for L perforated pins, 24.2% for LT perforated pins, and 29.2% for LTV perforated pins, indicating that LTV perforated pin fins are the lightest of the configurations studied.

6. The LTV pin-fin array, featuring vertical perforations, demonstrates superior thermal management capabilities due to enhanced turbulence and better mixing of the working fluid.

7. Fin effectiveness significantly improves with the introduction of perforations; specifically, for L, LT, and LTV perforated pin-fin arrays, it is 14, 34, and 57% higher, respectively, in comparison with solid pin fins.

8. LTV perforated pin-fins exhibit the highest reduction in surface temperatures compared to solid, L, and LT perforated pin-fins.

4.1 Future perspective

This study has clearly demonstrated that when considering solid pin-fin heat sinks against perforated ones, the former provides better thermal performance. The optimized LTV perforation configuration shows the greatest heat transfer efficiency and the least pressure drop, making it an ideal choice for thermal management in electronic devices. The significant weight reduction and lower operating temperatures further enhance the practicality of using perforated heat sinks in various industrial applications. These findings contribute valuable insights into the design of more efficient and compact cooling solutions, paving the way for advancements in electronic device thermal management. Future work could explore the integration of these perforation designs into commercial products and assess their long-term reliability under different operational stresses.

Nomenclature

Greek symbols

ΔP	Reduction in pressure within the test section (Pa)
ΔT	Temperature difference (°C)
ε_f	fin effectiveness
?	Enhancement ratio
η_f	fin efficiency
ρ_a	Density of air (kg m^{-3})
μ_a	Air's viscosity under dynamic conditions (kg $m^{-1}s^{-1}$)
ν_a	Viscosity in the kinematic realm of air (kg $m^{-1}s^{-1}$)

Subscripts

av	Average
b	Bulk
base	Refer to the fin base plate
f	Fin

in	Inlet
L	Longitudinal perforation
LT	Longitudinal and transverse perforation
LTV	Longitudinal, transverse, and vertical perforation
m	Mean
out	Outlet
s	Solid
∞	Free stream

Abbreviation

2D	Two dimension
3D	Three dimension
CFD	Computational fluid dynamics
f	Friction factor
h	Heat transfer coefficient
LT	Longitudinal/transverse perforation
LTV	Longitudinal/transverse/vertical perforation
Nu	Nusselt number
Re	Reynold number
R_{th}	Thermal resistance

Author details

Ndah Abdulrahman Alpha* and Aondoyila Kuhe
Department of Mechanical Engineering, Joseph Sarwuan Tarka University, Makurdi, Nigeria

*Address all correspondence to: alphazenga@gmail.com

References

[1] Chin S-B, Foo J-J, Lai Y-L, Yong TK-K. Forced convective heat transfer enhancement with perforated pin fins. Heat and Mass Transfer. 2013;**49**: 1447-1458

[2] Nafis BM, Whitt RA, Iradukunda C, Huitink D. Additive manufacturing for enhancing thermal dissipation in heat sink implementation: A review. Heat Transfer Engineering. 2021;**42**(12): 967-984

[3] Bisen VN, Sagar N. CFD analysis of heat transfer in annular fins of various profiles having different shapes of perforation. International Journal for Research in Applied Science and Engineering Technology. 2020;**8**(2020): 511-515

[4] Tijani AS, Jaffri NB. Thermal analysis of perforated pin-fins heat sink under forced convection condition. Procedia Manufacturing. 2018;**24**(2018):290-298

[5] Gupta D, Saha P, Roy S. Numerical investigation on heat transfer enhancement with perforated square micro-pin-fin heat sink for electronic cooling application. In: Proceeding of the IEEE RS/EPS/EDS Singapore Chapter, 21st Electronics Packaging Technology Conference (EPTC 2019). Singapore. 2019. DOI: 10.1109/EPTC47984.2019.9026623

[6] Adnan I, Khdair M. Numerical simulation of heat transfer of two-phase flow in mini-channel heat sink and investigation the effect of pin-fin shape on flow maldistribution. Engineering Analysis with Boundary Elements. 2023; **150**(2023):385-393. DOI: 10.1016/j.enganabound.2023.02.017

[7] Nilpueng K, Mehrdad M, Lazarus GA, Ahmet SD, Seon HA, Omid M, et al. Effect of pin fin configuration on thermal performance of plate pin fin heat sinks. Case Studies in Thermal Engineering. 2021;**27**(2021):101269

[8] Haque MR, Hridi TJ, Haque MM. CFD studies on thermal performance augmentation of heat sink using perforated twisted, and grooved pin fins. International Journal of Thermal Sciences. 2022;**182**:107832

[9] Alkhazaleh A, Fadi A, Mohamed YE, Selim B, Mathew M. Liquid cooling of microelectronic chips using MEMS heat sink: Thermohydraulic characteristics of wavy microchannels with pin-fins. International Journal of Thermofluids. 2023;**18**(2023):100313. DOI: 10.1016/j.ijft.2023.100313

[10] Wang D, Tao H. The parametric optimization to enhance the pin-fin microchannel heatsink performance using splitter and water/silver nanofluid applying the two-phase mixture model. Engineering Analysis with Boundary Elements. 2023;**146**(2023):216-225. DOI: 10.1016/j.enganabound.2022.10.026

[11] Ibrahim TK, Al-Sammarraie AT, Al-Jethelah MS, Al-Doori WH, Salimpour MR, Tao H. The impact of square shape perforations on the enhanced heat transfer from fins: Experimental and numerical study. International Journal of Thermal Sciences. 2020;**149**:106144

[12] Chingulpitak S, Ahn HS, Asirvatham LG, Wongwises S. Fluid flow and heat transfer characteristics of heat sinks with laterally perforated plate fins. International Journal of Heat and Mass Transfer. 2019;**2019**(138):293-303

[13] Kore SS, Yadav R, Chinchanikar S, Tipole P, Dhole V. Experimental

investigations of conical perforations on the thermal performance of cylindrical pin fin heat sink. International Journal of Ambient Energy. 2020;**43**(1):3431-3442. DOI: 10.1080/01430750.2020.1834451

[14] Gupta D, Saha P, Roy S. Computational analysis of perforation effect on the thermo-hydraulic performance of micro pin-fin heat sink. International Journal of Thermal Sciences. 2021;**2021**(163):106857

[15] Vishvas SC, Jagadale MR, Bhatkar VW. Heat transfer enhancement using perforated pin fins. International Journal for Technological Research in Engineering. 2015;**3**(2):2347-4718

[16] Ji-Jinn F, Pui S-Y, Lai Y-L, Chin S-B. Forced convective heat transfer enhancement with perforated pin fins subject to an impinging flow. SEGi Review. 2012;**5**(1):29-40

[17] Jaffal H, M, Jebur HS, Hussein AA. Numerical and experimental investigations on the performance characteristics for different shapes pin fin heat sink. International Journal of Computation and Applied Sciences IJOCAAS. 2018;**4**(3). ISSN: 2399-4509

[18] Kotcioglu I, Caliskan S, Baskaya S. Experimental study on the heat transfer and pressure drop of a cross-flow heat exchanger with different pin–fin arrays. Heat and Mass Transfer. 2011; **47**(9):1133

[19] Hwang SW, Kim DH, Min JK, Jeong JH. CFD analysis of fin tube heat exchanger with a pair of delta winglet vortex generators. Journal of Mechanical Science and Technology. 2012;**26**(2012): 2949-2958

[20] Koga A, Edson C, Nova H, Lima C, Silva E. Development of heat sink by using topology optimization. International Journal of Heat and Mass Transfer. 2013;**64**(2013):759-722

[21] Huang Y, Yang Q, Miao J, Tang K, Zhao J, Huang J, et al. Experimental investigation on flow boiling characteristics of a radial micro pin–fin heat sink for hotspot heat dissipation. Applied Thermal Engineering. 2023;**219** (Part C):119622. DOI: 10.1016/j.applthermaleng.2022.119622

[22] Al-Damook A, Summers J, Kapur N, Thompson H. Effect of temperature dependent air properties on the accuracy of numerical simulations of thermal airflows over pinned heat sinks. International Communications in Heat and Mass Transfer. 2016;**78**:163-167

[23] Maji A, Bhanja D, Patowari PK, Numerical investigation on heat transfer enhancement of heat sink using perforated pin fins with inline and staggered arrangement, Applied Thermal Engineering. 2017;**125**(2017): 596-616. ISSN 1359 4311. DOI: 10.1016/j.applthermaleng.2017.07.053

[24] Wu M, Shen S, Yang X, Dong W, Song F, Zhu Y, et al. Advances in the enhancement of bionic fractal microchannel heat transfer process. Journal of Thermal Analysis and Calorimetry. 2023;**148**:13497-13517. DOI: 10.1007/s10973-023-12620-4

[25] Cengel YA, Cimbala JM, Kanoğlu M. Fluid mechanics. In: Fundamentals and Applications. New York: McGraw-Hill; 2006

[26] Al-Damook A, Kapur N, Summers JL, Thompson HM. An experimental and computational investigation of thermal air flows through perforated pin heat sinks. Applied Thermal Engineering. 2015;**89**: 365-376

[27] Holman JP. Heat Transfer. 10th ed. New York: McGraw-Hill; 2010. ISBN 978–0–07-352936-3

[28] Mehedi E, Mohammad A, Islam MQ, Muhsia T. Thermal and hydraulic performance analysis of rectangular fin arrays with perforation size and number. Procedia Engineering. 2015;**105**: 184-191

[29] Hasanpour A, Farhadi M, Sedighi K. A review study on twisted tape inserts on turbulent flow heat exchangers: The overall enhancement ratio criteria. International Communications in Heat and Mass Transfer. 2014;**55**:53-62

[30] Shaeri MR, Yaghoubi M, Jafarpur K. Heat transfer analysis of lateral perforated fin heat sinks. Applied Energy. 2009;**86**(2009):2019-2029

[31] Chamoli S, Chauhan R, Thakur N. Numerical analysis of heat transfer and thermal performance analysis of surface with circular profile fins. International Journal of Energy Science. 2011;**1**(2011): 11-18

[32] Rao Y, Wana C, Xu Y. An experimental study of pressure loss and heat transfer in the pin fin-dimple channels with various dimple depths. International Journal of Heat and Mass Transfer. 2012;**55**(2012):6723-6733

[33] Incropera FP, DeWitt DP, Bergman TL, Lavine AS. Fundamental of Heat and Mass Transfer. 6th ed. New York: John Wiley and Sons Inc.; 2007

[34] Abdulrahman Alpha N, Aondover Iortyer Humphrey A, Aondoyila K. Experimental and numerical studies of the effect of perforation configuration on heat transfer enhancement of pin fins heat sink. Heat Transfer. 2024;**53**(5): 2525-2555. DOI: 10.1002/htj.23051

[35] Al-Sallami W, Al-Damook A, Thompson HM. A numerical investigation of the thermal-hydraulic characteristics of perforated plate fin heat sinks. International Journal of Thermal Sciences. 2017;**121**:266-277

[36] Alam T, Saini RP, Saini JS. Heat and flow characteristics of air heater ducts provided with turbulators—A review. Renewable and Sustainable Energy Reviews. 2014;**31**:289-304

[37] Gurrum SP, Suman SK, Joshi YK, Fedorov AG. Thermal issues in next-generation integrated circuits. IEEE Transactions on Device and Materials Reliability. 2004;**4**:709-714

[38] Yuan W, Zhao J, Tso CP, Wu T, Liu W, Ming T. Numerical simulation of the thermal hydraulic performance of a plate pin fin heat sink. Applied Thermal Engineering. 2012;**48**:81-88

[39] Sara ON, Pekdemir SY, Yilmaz M. Heat transfer enhancement in a channel flow with perforated rectangular blocks. International Journal of Heat and Fluid Flow. 2001;**22**:509-518

[40] Sara ON, Pekdemir T, Yapici S, Ersahan H. Thermal performance analysis for solid and perforated blocks attached on a flat surface in duct flow. Energy Conversion & Management. 2000;**41**:1010-1028

[41] Ismail MF, Reza MO, Zobaer MA, Ali M. Numerical investigation of turbulent heat convection from solid and longitudinally perforated rectangular fins. Procedia Engineering. 2013;**56**: 497-502

[42] Noaman S, Khan R, Faizanurrahman SA, Ansari T, Tarique K. Heat transfer enhancement through perforated fin. IOSR Journal of

Mechanical and Civil Engineering (IOSR-JMCE). e-ISSN: 2278-1684,p-ISSN: 2320-334X. pp. 72-78. 6th National Conference RDME 2017, 17th- 18th March 2017. Pune: M.E.S. College of Engineering; 2017. p. 411001. DOI: 10.9790/1684-17010047278

[43] Shaeri MR, Yaghoubi M. Numerical analysis of turbulent convection heat transfers from an array of perforated fins. International Journal of Heat and Fluid Flow. 2009;**30**:218-228

[44] Shaeri MR, Jen TC. The effects of perforation sizes on laminar heat transfer characteristics of an array of perforated fins. Energy Conversion and Management. 2012;**64**:328-334

Chapter 4

Thermal Conductance of Nanofluids and Effective Mechanisms: A Review

Elif Begum Elcioglu

Abstract

Nanofluids, as combinations of nanoparticles, base fluids, and other optional additives, are innovative materials for thermal engineering. Nanofluids potential comes mainly from their enhanced and tuneable thermal conductivities, which, in turn, result in important advantages in heat transfer applications. It is critically important to understand how nanofluids thermal conductivity can be tuned and improved as necessitated by the application, in order for the potential of nanofluids can be realized at a large scale. There has been a significant debate on whether the thermal conductivity improvement with nanofluids has explicable and quantifiable physical/chemical mechanism so that optimal thermal nanofluids and nanofluids for extreme thermal conditions could be developed. So far, no universal mechanism has shone out as being responsible for thermal conductivity increases, and further, such increases have ceased to be sustainable. In this chapter, a systematic literature review on nanofluids thermal conductivity is provided by focusing on nanofluids (component type, particle morphology, concentration, etc.) and process parameters (temperature, ultrasonication). The mechanisms that may contribute to nanofluids thermal conductivity improvement are summarized. The chapter intends to provide the common and uncommon trends and results from experimental and theoretical research and present an overview of thermal nanofluids effective mechanisms.

Keywords: nanofluids, nanoparticles, dispersions, thermal conductivity, heat transfer

1. Introduction

Thermal conductivity is a material-specific transport property and it is a measure of how easily heat is transported through a medium. As technological advancements speed up the miniaturization trend and smaller functioning units are packed together, thermal management becomes a critical aspect. Lithium-ion batteries operating temperature being in the range of –20 to 60°C for discharging and 0–55°C during charging makes thermal management crucial for high efficiency and safe usage [1]. Liquid cooling of batteries with conventional fluids, as water may cause corrosion and oils may be inflammable, and both being of low thermal conductivity is challenging [2]. Data centers, with their high energy needs for data traffic, storage, and processing, use approximately 38% of their energy supply for cooling [3], and they require effective cooling systems. Photovoltaic (PV) systems also require the operating temperature to be as low as possible, and since incorporating passive methods results in non-uniform surface temperatures [4], sophisticated liquid cooling systems are required. Although the transferred heat can be increased by optimizing system geometry and choosing components that facilitate heat removal, the results of such efforts are limited by the heat transfer capabilities of the chosen fluids. Typically, pure fluids have low thermal conductivities [5], and when properly combined with solid nanoparticles, they can be fine-tuned to exhibit performances for enhanced thermal management and heat transfer.

Nanofluids as dispersions of nanoparticles in industrial fluids, after being named by Choi and Eastman [6] in 1995, have attracted considerable interest in the research community due to their potential and outstanding characteristics that are tuneable. These characteristics include thermal conductivity (as elaborated on in this chapter), viscosity [7], specific heat [8], optical properties [9, 10], surface tension [11, 12], and contact angle [13], making them adaptable to many technological applications. Due to the fact that nanofluids have significant usage potential in a variety of applications, there are quite a high number of papers that focused on applications of nanofluids, for example, in solar collectors [14–16], photovoltaic/thermal systems [17, 18], fuel cells [19], thermal processing of food products [20], electronics cooling [21], minimum quantity lubrication [22], spray cooling [23], turbulent heat transfer [24], and natural convection systems [25, 26], among others. The nanofluids heat transfer literature has mainly been driven by their thermal conductivity, enhancements which have sometimes been regarded as anomalous [27]. For example, Lin et al. [28] concluded that the presence of graphene nanoparticles could result in a thermal conductivity increase of 1.2 to 83.4% compared to the base fluid. On the other hand, some research reported otherwise, stating there were no anomalous thermal conductivity enhancement exhibited by their nanofluid samples [29].

An important question is what made nanofluids transition into industrial products challenging and why nanofluids had not replaced conventional fluids yet? There have been evaluations assessing the challenges and complexities brought by nanofluids [21, 30, 31] along with assessments on environmental and socio-economic aspects regarding their usage [32–38] in different scales. From thermal engineering point of view, understanding thermal conductivity of nanofluids and effective mechanisms on their thermal characteristics, together with specific heat capacity, density, and viscosity are of key importance in development of sustainable and realistic strategies for their transition into industrial heat transfer fluids. In this Chapter, the recent literature on nanofluids thermal conductivity is reviewed and discussions based on nanofluid components as well as preparation conditions are provided. Specific attention is given to the mechanisms that were proposed to be effective on the thermal conductivity trends and behaviors of nanofluids.

2. Nanofluids thermal conductivity

Addition of solids into liquids has become more effective with nanoparticle technology since sub-micron particles can be better dispersed in liquids compared to larger and heavier microparticles due to gravity effect. In addition, enhancing liquids with nanoparticles bring about new functionalities, as at the nanoscale, properties of materials become size-dependent, and it is possible for materials to exhibit properties different than their bulk counterparts. Masuda et al. [39] in 1993 reported on the thermal conductivities of water-based Al_2O_3, SiO_2, and TiO_2 nanofluids at temperatures up to 340 K. Their results showed that the thermal conductivities of water-based Al_2O_3 and TiO_2 nanofluids were higher than that of water while the thermal conductivity of water- SiO_2 nanofluid did not show any increase. After that, in 1995, Choi and Eastman [6] conducted a theoretical work on the thermal conductivity of copper nanoparticles dispersed in water, and they reported that nanofluids with their thermal conductivity could be the best hope for enhanced heat transfer. Later on, Eastman et al. [40] studied pump oil-based Cu as well as water-based Al_2O_3 and CuO nanofluids thermal conductivity for up to 5% concentration in volume. Their results revealed around 60% and more increase in thermal conductivity for 5% concentration. In another work, Eastman et al. [41] studied ethylene-glycol based Cu nanofluids and their results showed that the effective thermal conductivity increased by up to 40% for 0.3 vol.% Cu nanoparticles. This research formed the basis of nanofluids development as heat transfer fluids and accelerated the research communities focus in this area.

2.1 Overview of recent literature

Thermal conductivity is one of the defining factors of nanofluids potential as heat transfer fluids. Enhanced thermal conductivity positively affects the heat transfer coefficient, which in turn enhances the efficiency of cooling and thermal management procedures. In this regard, performance of different types of nanofluids was investigated theoretically and experimentally in various systems and applications, including cooling of lithium-ion batteries [1, 2], cooling of data centers [3], GPU cooling [42], cooling of truck engines [43], cooling of electronic devices [44], minimum quantity lubrication during machining processes such as grinding [45] and lathe machining [46], and cooling of nuclear reactors [47, 48] among others.

Since thermal conductivity has been one of the main characteristics of nanofluids making them promising in energy applications, and that by nature nanofluids are tuneable materials, it is important to understand the mechanisms and effects involved in their thermal conductivity. The mechanisms considered or proposed to be effective on nanofluids thermal conductivity include Brownian motion [28, 49], thermal boundary resistance (Kapitza resistance) [50, 51], mass difference scattering [50], thermophoresis [28, 49] and osmophoresis [49], nanoparticle aggregation [28, 51, 52], liquid layering [28] or interfacial layer [53], ballistic transport [28], and near-field thermal radiation [28, 54, 55]. It is also known that the method of choice for thermal conductivity measurement (e.g., transient hot wire (THW) method, temperature oscillation method, steady-state parallel plate method [56], 3ω method [57, 58], laser flash method [56], and transient plane source method [56]) affects the measured data. For example, Buongiorno et al. [59] considered different measurement techniques for the thermal conductivity of nanofluids and concluded that the data were indicative of clear differences when measurement technique was changed, while such differences were less pronounced once the data were normalized using base fluid thermal conductivity. Apart from the mentioned mechanisms and reasons, the parameters that are effective on nanofluid thermal conductivity can be classified as directly tuneable (controllable and quantifiable) and not directly tuneable. The tuneable parameters include nanoparticle and base fluid type, their relative amounts, presence of stabilizing additives, choice of applying ultrasonication, if applied, type of ultrasonication, pH value, and temperature.

2.1.1 Effect of nanoparticle concentration

The effect of nanoparticle concentration on thermal conductivity and thermal conductivity augmentation has been among the most well-understood mechanisms. Since the thermal conductivity of most of the solid nanoparticles is well above that of the base fluids, there is a positive correlation between concentration and thermal conductivity for such cases. Esfahani and Toghraie [60] studied the thermal conductivity of SiO_2–water and ethylene glycol (40:60) nanofluids of 0.1–5 vol.% at 25–50°C using THW method. Their results showed that for a 5% nanoparticle volume fraction at 50°C, the thermal conductivity of the nanofluid increased by 50.9% relative to the base fluid at 25°C. They concluded that the increase of nanoparticle concentration at higher temperatures had a more significant effect on nanofluid's conductivity. Xie et al. [61] studied ethylene glycol-based MgO, TiO_2, ZnO, Al_2O_3, and SiO_2 nanofluids up to 5% volume concentration using the THW method. Their results showed that at a 5% concentration for MgO-ethylene glycol nanofluid, a thermal conductivity enhancement of 40.6% was obtained, and the thermal conductivity enhancement rate decreased as the nanoparticle concentration increased. On the other hand, depending on the thermal conductivity of the nanoparticles, the thermal conductivity of the nanofluid may decrease. Hwang et al. [62] studied thermal conductivity of water, ethylene glycol, and oil-based multi-walled carbon nanotube (MWCNT), CuO, SiO_2, and fullerene nanofluids via THW method. Their results showed that thermal conductivity increased with increasing MWCNT, CuO, and SiO_2 concentration while it decreased with increased fullerene concentration due to the fact that fullerene had a thermal conductivity that was lower (i.e., 0.4 W/mK) than that of the base fluid. Nanoparticle concentration effect on thermal conductivity is clearer than the effects of other parameters. On the other hand, effect of concentration is multifaceted, and should be considered, focusing on different concentration ranges. For example, for dilute nanofluids, thermal conductivity is rather independent of particle-particle interactions, while importance of particle-particle interactions increase at higher concentrations [49]. Increased particle concentration, although it supports improved thermal conductivities, can make it more challenging to prepare a colloidally stable dispersion.

2.1.2 Effect of nanoparticle morphology

Nanoparticles physical characteristics such as their size, shape, surface, and dimensionality as 0D to 3D are parameters of their morphology, and these parameters have a direct effect on nanoparticle properties including thermal conductivity. When it comes to nanofluids; nanoparticle size, shape, and aspect ratio have an effect on their colloidal stability, thereby both directly and indirectly affecting thermal conductivity. Darvanjooghi and Esfahany [63] studied the thermal conductivity of ethanol based SiO_2 nanofluids with a particle size of 10.6, 20, 38.6 and 62 nm via THW method. Their results showed that increasing particle size increased the thermal conductivity. The effect of particle size was more pronounced for sizes <20 nm while for >20 nm particle size increase resulted in an increase of the thermal conductivity slightly. Maheshwary et al. [64] measured the thermal conductivity of water-based TiO_2 nanofluids of 0.5–2.5% weight concentration between 303 and 353 K. Their results showed that the thermal conductivity was higher for smaller particle sizes (31.339, 30.381, 20.657, and 16.312 nm) attained via ultrasonication process. Particle shape (spherical, cubic, and rod) was also shown to be effective on thermal conductivity. Turgut et al. [65] studied the thermal conductivity of Al_2O_3-water nanofluids of 1-6.33% volume concentration for particle sizes of 10 and 30 nm at 25°C using 3ω method. Their results showed that the thermal conductivity enhancement rate was lower than 1%, even for the highest concentration, and thermal conductivity was unchanging with particle size. Ali et al. [66] studied water-based Al_2O_3 nanofluids with nanoparticle diameters of 11, 25, and 63 nm with the CTAB added as a surfactant. The thermal conductivity measurements performed via the THW method revealed that thermal conductivity increased as nanoparticle size increased. Timofeeva et al. [67] studied Al_2O_3-water:ethylene glycol (1:1) nanofluids thermal conductivity by considering Al_2O_3 nanoparticle shapes of platelets, blades, rods, and bricks, for which the nanoparticle concentrations up to around 8% were considered. The thermal conductivity enhancements were the highest for dispersions containing blade shaped nanoparticles, followed by platelets, cylinders, and bricks. Jeong et al. [68] studied ZnO-water nanofluids thermal conductivity by considering spherical and near-rectangular shapes of ZnO nanoparticles. Their results showed that nanofluids with near-rectangular shaped nanoparticles exhibited higher thermal conductivities compared to those with spherical-shaped nanoparticles, which was attributed to the differences in effective aggregate radii.

Huang et al. [69] studied the thermal conductivity of Al_2O_3-water nanofluids using molecular dynamics in a volume fraction range of 1.24–6.2% between 290 and 360 K for different sphericity values in the range 0.6975-1 while a spherical particle has a sphericity value of 1. Their results showed that the thermal conductivity was most sensitive to temperature, followed by sphericity and nanoparticle concentration, and thermal conductivity decreased with increased sphericity. Jin et al. [70] studied the effect of the interfacial layer on the thermal conductivity of nanofluids containing core-shell nanoparticles using molecular dynamics. Their results showed that nanofluids with core-shell nanoparticles showed higher thermal conductivity enhancement compared to those with ordinary nanoparticles. There was an optimum core-shell ratio for the maximum thermal conductivity.

2.1.3 Effect of nanoparticle aggregation

Consideration of additives when formulating nanofluids is of critical importance in sustaining colloidal stability and improvements in nanofluid characteristics. Nanoparticles tend to come together and form agglomerates [71], aggregates, and clusters, resulting in larger particles that mostly experience sedimentation [72]. This mechanism has multiple effects including increase of particle size, change in property values [73], and change in particle morphology. **Figure 1** depicts different states of particles in a dispersion.

Timofeeva et al. [29] studied Al_2O_3-water and Al_2O_3-ethylene glycol nanofluids thermal conductivity, and stated that the discrepancy between the available thermal conductivity enhancement results would have been due to the differences of

Figure 1.
Particles in dispersions. From left to right: monodispersed nanoparticles, agglomerates, aggregates, surface functionalized nanoparticles, core-shell nanoparticles (images are not to scale).

agglomerate sizes and shapes. Wei et al. [74] developed a model for nanofluid thermal conductivity that depends on fractal distribution characteristics of nanoparticle aggregation and concentration. Wang et al. [75] studied the impact of aggregation on Cu-Ar nanofluid thermal conductivity via molecular dynamics simulations. Their results showed that the thermal conductivity increased as the fractal dimension of aggregation decreased. Cui et al. [76] studied Cu-Ar nanofluids heat transfer augmentation via molecular dynamics. They focused on the migration and rotation of nanoparticles in the base fluid, and concluded that the effect of nanoparticle rotation was comparable to the effect of translational movement on the heat transfer enhancement. Evans et al. [51] concluded that the cluster morphology was an effective factor on thermal conductivity, and that the thermal conductivity of nanofluids and nanocomposites was positively correlated to nanoparticle clustering. Wang et al. [77] studied thermal conductivity of dilute Cu-Ar nanofluids using molecular dynamics, and concluded that the contribution of the interfacial layer to the thermal conductivity of nanofluids could be neglected. Khodayari et al. [78] studied interfacial thermal resistance (Kapitza resistance) and liquid layering (nanolayer) effects on alumina-water nanofluids thermal conductivity. Their results showed that nanolayer had very little effect on thermal conductivity while further research was needed to accurately measure the Kapitza resistance effect at the solid-liquid interface. Du et al. [79] used multi-particle collision dynamics and molecular dynamics simulations to evaluate the impact of aggregation morphology on Cu-water nanofluids thermal conductivity. Their results revealed that at a fixed nanoparticle size and concentration, thermal conductivity increased as the aggregate fractal dimension decreased.

2.1.4 *Effect of additives and ultrasonication*

While thermal conductivity of nanofluids has been one of the reasons why they are considered promising as heat transfer fluids, sustaining of nanofluid stability being challenging has been among the top reasons why nanofluids have not so far replaced conventional heat transfer fluids. For example, Xuan et al. [80] investigated the impinging heat transfer performance of Cu-water nanofluids in the presence and absence of SDBS as a surfactant, and their results revealed that the presence of SDBS decreased the heat transfer coefficient at a fixed nanoparticle concentration. Das et al. [81] investigated the thermal conductivity of aqueous Al_2O_3 nanofluids with the aim of observing the effects of CTAB, SDBS, and SDS surfactants using THW method. Among these surfactants, SDBS showed better performance against CTAB and SDS. Their results showed that the addition of SDBS slightly decreased the thermal conductivity of the nanofluid, but this decrease was smaller than the measurement uncertainty. Khairul et al. [82] studied water-based CuO and Al_2O_3 nanofluids thermal conductivity using THW method. The samples were stabilized using SDBS. Their results showed that there was an optimal SDBS concentration for which the thermal conductivity ratio was maximized. The addition of surfactant beyond this optimal limit caused the thermal conductivity ratio to decrease. Asadi et al. [83] studied thermal conductivity of $Mg(OH)_2$-water nanofluids of 0.1–2% concentration at varying temperatures of 25–50°C and formulated their samples by adding CTAB, SDS, or Oleic Acid as surfactants. Thermal conductivity measurements were done via THW method. The samples containing CTAB showed the best stability, and the authors stated that the sample with surfactant would have higher thermal conductivity compared to the sample without surfactant. In the work of Asadi et al. [83] on the thermal conductivity of $Mg(OH)_2$-water nanofluids stabilized using CTAB surfactant, effects of the ultrasonication duration (10–160 min) were also investigated. Results showed that ultrasonication for 30 minutes resulted in the highest stability. Mahbubul et al. [84] studied 0.5 vol.% Al_2O_3-water nanofluids thermal conductivity depending on ultrasonication process for 0.5 to 5 hours. Their results showed that in the range of 10–50°C, thermal conductivity decreased with ultrasonication in the range of 0–1 hour and exhibited a continuously increasing trend for 1–5 hours of ultrasonication. Literature shows that the amount of additives [85] as well as the pH value [29, 85, 86] of nanofluids should be optimized to exhibit high thermal conductivity.

2.2 Effective mechanisms explaining nanofluids thermal conductivity

By parametric investigations on nanofluid thermal conductivity, several mechanisms were brought to the fore as effective mechanisms. Understanding of the mechanisms that are effective on thermal conductivity and thermal conductivity enhancement is critically important to designing thermal nanofluids. Temperature as a process parameter has a significant effect on nanofluids thermal conductivity, in addition to the fact that temperature has sole effect as well as interaction effects with other parameters (particle concentration, particle size, agglomeration, and so on). Mariano et al. [87] studied thermal conductivity of Co_3O_4-ethylene glycol nanofluids of different concentrations using THW between 283.15 and 323.15 K. Their results showed a slight decrease of nanofluid thermal conductivity with temperature, while the base fluid thermal conductivity was increasing with increasing temperature. Yang and Han [88] studied the thermal conductivity of Bi_2Te_3 nanorods nanofluids of 20 nm diameter and 170 nm length, and concluded that the enhancement in thermal conductivity was decreasing with temperature.

In the literature, there are many reports available stating that nanofluid thermal conductivity increases as temperature increases [63, 64, 81, 83, 89–91] and nanoparticle concentration increases [63, 64, 81–83, 89–91]. On the other hand, nanofluid thermal conductivity can decrease with nanoparticle concentration, when the nanofluid is formulated with a low thermal conductivity nanoparticle such as fullerene [62]. The effect of the particle size is not clear, as reports are available to show positive [63, 66] and negative [92, 93] non-monotonous correlation of nanofluid thermal conductivity with particle size. Apart from the parameters that are changeable and tuneable by the researchers, there are other effects some of which are not necessarily directly measurable but present. **Figure 2** classifies such mechanisms and parameters for designing optimal nanofluids for heat transfer.

Esfahani and Toghraie [60] discussed some effective mechanisms for nanofluid thermal conductivity. They attributed the increase of thermal conductivity at higher temperatures to an increased number of collisions and the combined effect of temperature and concentration to the fact that the intermolecular bonds in the fluid layers were loosened, along with the increased number of collisions and Brownian motion. Darvanjooghi and Esfahany [63] attributed the increase in thermal conductivity ratio of nanofluid with temperature to micro-convection effect. Darvanjooghi and Esfahany [63] reported by keeping the agglomeration minimum that the thermal conductivity enhancement with particle size was not due to agglomeration, but rather it was attributed to the surface structure and thereby to the interfacial thermal resistance reduction. Das et al. [81] attributed the increase of thermal conductivity with increasing temperature to increased importance of Brownian motion, van der Waals, and electrostatic forces.

Parameters tuneable by specialists	Parameters not directly tuneable
• Nanoparticle type	• Brownian motion
• Nanoparticle size, shape, aspect ratio, concentration	• Micro-convection
• Base fluid type	• Agglomeration
• Use of surfactants	• Ballistic phonon transport
• Use of surface modification	• Thermal boundary resistance
• Use of ultrasonication	• Interfacial nanolayer
• Temperature	• Thermophoresis
• pH value	• Near-field thermal radiation
• Measurement technique	• Osmophoresis
	• Mass difference scattering

Figure 2.
Nanofluid thermal conductivity parameters.

Agarwal et al. [89] also concluded Brownian motion to be the main mechanism behind the thermal conductivity enhancement of Al_2O_3 nanofluids, along with the formation of a solid-like layer at the solid-liquid interface. Koo and Kleinstreuer [49] considered CuO-water nanofluids thermal conductivity and reported that on the effective thermal conductivity, the effect of Brownian motion was more pronounced compared to those of thermophoretic and osmophoretic effects. Song et al. [94] focused on liquid phonon theory and stated that higher nanoparticle conductivity did not automatically result in higher nanofluid thermal conductivity, and the phonon mismatch between liquid and nanoparticle must be taken into consideration. Iacobazzi et al. [28] compared the impacts of mass difference scattering, nanolayer, Brownian motion, ballistic phonon motion, and Kapitza resistance, and concluded that mass difference scattering reduces nanofluid thermal conductivity.

3. Conclusions

In this Chapter, a review of nanofluid thermal conductivity field focused on process and nanofluid parameter effects is provided. For this purpose, a systematic literature review is conducted on the experimental and theoretical research sorted by their relevance and recency, although some fundamental works dated earlier are also mentioned for the sake of completeness. Literature shows that some parameters have undeniable impacts on nanofluids thermal conductivity. For example, addition of nanoparticles in a base fluid alters the thermal conductivity of the nanofluid, depending on the thermal conductivity of the nanoparticle, and this trend is augmented by the nanoparticle concentration. On the other hand, there is a limit for which the effect of dispersed nanoparticle amount is straightforward, since increased concentration means increased tendency of nanoparticles to collide and form clusters, which brings the nanoparticle agglomeration effect on nanofluids thermal conductivity. The effect of agglomerate size and shape is not straightforward, and since it is challenging to control agglomerate size experimentally, molecular dynamics-based works are focused on this phenomenon more. When nanoparticles are agglomerated, ways to improve nanoparticles dispersion and make nanoparticle size range as small as possible are used, including stirring of the dispersion and ultrasonication process.

Although stirring of the solution does not directly affect the particle size distribution and hence nanofluid properties, ultrasonication has a profound effect on thermal conductivity. The literature reveals that there is an optimum condition for the ultrasonication process that results in maximum thermal conductivity. The same result applies to the use of surfactants, as there is an optimum type and concentration for the stabilizers for the thermal conductivity to be maximized. The need for optimizing ultrasonication and stabilizing additives directly means that the dispersion state has a major impact on the thermal conductivity of nanofluids. In order to design thermal nanofluids, thermal conductivity is an indispensable characteristic that directly affects the heat transfer coefficient. On the other hand, it is thermal engineers' duty to consider the enhanced viscosity of nanofluids compared to the base fluids and the associated increases in pump power requirements. In addition to the viscosity-related expenses, it is necessary to develop nanofluids in a way that the fabrication routes are environment-friendly and the usage, handling, and recycling procedures are well-thought.

Conflict of interest

The author declares no conflict of interest.

Author details

Elif Begum Elcioglu
Faculty of Engineering, Department of Mechanical Engineering, Eskisehir Technical University, Eskisehir, Turkey

*Address all correspondence to: ebelcioglu@eskisehir.edu.tr

References

[1] Haddad Z, Belkadi D, Mourad A, Aissa A, Said Z, Younis O, et al. Advancements and comprehensive overview of thermal management systems for lithium-ion batteries: Nanofluids and phase change materials approaches. Journal of Power Sources [Internet]. 2024;**603**:234382. DOI: 10.1016/j.jpowsour.2024.234382

[2] Xu X, Weng K, Lu X, Zhang Y, Zhu S, Zou D. Functional thermal fluids and their applications in battery thermal management: A comprehensive review. Journal of Energy Chemistry [Internet]. 2024;**94**:78-101. DOI: 10.1016/j.jechem.2024.02.054

[3] Amorim GS, Belman-Flores JM, Mendes R, de Sandoval OR, Khosravi A, Garcia-Pabon JJ. Recent advancements in thermal management technologies for cooling of data centers. Journal of the Brazilian Society of Mechanical Sciences and Engineering [Internet]. 2024;**46**:472. DOI: 10.1007/s40430-024-05048-w

[4] Javidan M, Moghadam AJ. Effective cooling of a photovoltaic module using jet-impingement array and nanofluid coolant. The International Communications in Heat and Mass Transfer [Internet]. 2022;**137**:106310. DOI: 10.1016/j.icheatmasstransfer.2022.106310

[5] Kakaç S, Pramuanjaroenkij A. Review of convective heat transfer enhancement with nanofluids. International Journal of Heat and Mass Transfer [Internet]. 2009;**52**(13-14): 3187-3196. DOI: 10.1016/j.ijheatmasstransfer.2009.02.006

[6] Choi SUS, Eastman JA. Enhancing thermal conductivity of fluids with nanoparticles. In: Siginer DA, Wang HP, editors. Developments and Applications of Non-Newtonian Flows, FED. Vols. 231/MD vol. 66, New York: ASME; 1995. pp. 99-103

[7] Mahbubul IM, Saidur R, Amalina MA. Latest developments on the viscosity of nanofluids. International Journal of Heat and Mass Transfer. 2012;**55**:874-885

[8] Shahrul IM, Mahbubul IM, Khaleduzzaman SS, Saidur R, Sabri MFM. A comparative review on the specific heat of nanofluids for energy perspective. Renewable and Sustainable Energy Reviews. 2014;**38**:88-98

[9] Sharaf OZ, Rizk N, Joshi CP, Abi Jaoudé M, Al-Khateeb AN, Kyritsis DC, et al. Ultrastable plasmonic nanofluids in optimized direct absorption solar collectors. Energy Conversion and Management [Internet]. 2019;**199**(July):112010. DOI: 10.1016/j.enconman.2019.112010

[10] Li X, Zeng G, Lei X. The stability, optical properties and solar-thermal conversion performance of SiC-MWCNTs hybrid nanofluids for the direct absorption solar collector (DASC) application. Solar Energy Materials & Solar Cells. 2020;**206**(1):110323

[11] Elcioglu EB, Murshed SMS. Ultrasonically tuned surface tension and nano-film formation of aqueous ZnO nanofluids. Ultrasonics Sonochemistry. 2020;**72**:105424

[12] Su G, Yang L, Liu S, Song J, Jiang W, Jin X. Review on factors affecting nanofluids surface tension and mechanism analysis. Journal of Molecular Liquids [Internet].

2024;**407**:125159. DOI: 10.1016/j.molliq.2024.125159

[13] Hernaiz M, Alonso V, Estellé P, Wu Z, Sundén B, Doretti L, et al. The contact angle of nanofluids as thermophysical property. Journal of Colloid and Interface Science. 2019;**547**:393-406

[14] Alawi OA, Kamar HM, Mallah AR, Mohammed HA, Kazi SN, Sidik NAC, et al. Nanofluids for flat plate solar collectors: Fundamentals and applications. Journal of Cleaner Production. 2021;**291**:125725

[15] Muhammad MJ, Muhammad IA, Che Sidik NA, Muhammad Yazid MNAW. Thermal performance enhancement of flat-plate and evacuated tube solar collectors using nanofluid: A review. International Communications in Heat and Mass Transfer. 2016;**76**:6-15

[16] Rasih RA, Sidik NAC, Samion S. Recent progress on concentrating direct absorption solar collector using nanofluids: A review. Journal of Thermal Analysis and Calorimetry. 2019;**137**(3):903-922

[17] Das D, Kalita P, Roy O. Flat plate hybrid photovoltaic- thermal (PV/T) system: A review on design and development. Renewable and Sustainable Energy Reviews [Internet]. 2018;**84**:111-130. DOI: 10.1016/j.rser.2018.01.002

[18] Hemmat Esfe M, Kamyab MH, Valadkhani M. Application of nanofluids and fluids in photovoltaic thermal system: An updated review. Solar Energy [Internet]. 2020;**199**:796-818. DOI: 10.1016/j.solener.2020.01.015

[19] Sofiah AGN, Pasupuleti J, Samykano M, Pandey AK, Rajamony RK, Sulaiman NF, et al. A class of promising fuel cell performance: International status on the application of nanofluids for thermal management systems. Materials Today Sustainability [Internet]. 2024;**26**:100709. DOI: 10.1016/j.mtsust.2024.100709

[20] Salari S, Jafari SM. Application of nanofluids for thermal processing of food products. Trends in Food Science and Technology [Internet]. 2020;**97**:100-113. DOI: 10.1016/j.tifs.2020.01.004

[21] Aglawe KR, Yadav RK, Thool SB. Preparation, applications and challenges of nanofluids in electronic cooling: A systematic review. Materials Today: Proceedings [Internet]. 2020;**43**:366-372. DOI: 10.1016/j.matpr.2020.11.679

[22] Chinchanikar S, Kore SS, Hujare P. A review on nanofluids in minimum quantity lubrication machining. Journal of Manufacturing Processes [Internet]. 2021;**68**:56-70. DOI: 10.1016/j.jmapro.2021.05.028

[23] Sanches M, Marseglia G, Ribeiro APC, Moreira ALN, Moita AS. Nanofluids characterization for spray cooling applications. Symmetry (Basel). 2021;**13**(5):1-19

[24] Yu W, France DM, Timofeeva EV, Singh D, Routbort JL. Comparative review of turbulent heat transfer of nanofluids. International Journal of Heat and Mass Transfer [Internet]. 2012;**55**(21-22):5380-5396. DOI: 10.1016/j.ijheatmasstransfer.2012.06.034

[25] Rahimi A, Saee AD, Kasaeipoor A, Malekshah EH. A comprehensive review on natural convection flow and heat transfer: The most practical geometries for engineering applications. International Journal of Numerical Methods for Heat and Fluid Flow. 2019;**29**(3):834-877

[26] Doganay S, Turgut A. Enhanced effectiveness of nanofluid based

natural circulation mini loop. Applied Thermal Engineering [Internet]. 2015;**75**:669-676. DOI: 10.1016/j.applthermaleng.2014.10.083

[27] Sergis A, Hardalupas Y. Anomalous heat transfer modes of nanofluids: A review based on statistical analysis. Nanoscale Research Letters. 2011;**6**:1-37

[28] Lin H, Jian Q, Bai X, Li D, Huang Z, Huang W, et al. Recent advances in thermal conductivity and thermal applications of graphene and its derivatives nanofluids. Applied Thermal Engineering [Internet]. 2023;**218**:119176. DOI: 10.1016/j.applthermaleng.2022.119176

[29] Timofeeva EV, Gavrilov AN, McCloskey JM, Tolmachev YV, Sprunt S, Lopatina LM, et al. Thermal conductivity and particle agglomeration in alumina nanofluids: Experiment and theory. Physical Review E. 2007;**76**:061203

[30] Sidik NAC, Mohammed HA, Alawi OA, Samion S. A review on preparation methods and challenges of nanofluids. International Communications in Heat and Mass Transfer. 2014;**54**:115-125

[31] Sukarno DH. Challenges for nanofluid applications in heat transfer technology. Journal of Physics Conference Series. 2017;**795**(012020):1-6

[32] Brignon JM. Socio-economic analysis: A tool for assessing the potential of nanotechnologies. Journal of Physics Conference Series. 2011;**304**(012069):1-8

[33] Otanicar TP, Golden JS. Comparative environmental and economic analysis of conventional and nanofluid solar hot water technologies. Environmental Science & Technology. 2009;**43**:6082-6087

[34] Faizal M, Saidur R, Mekhilef S, Alim MA. Energy, economic and environmental analysis of metal oxides nanofluid for flat-plate solar collector. Energy Conversion and Management [Internet]. 2013;**76**:162-168. DOI: 10.1016/j.enconman.2013.07.038

[35] Hernandez, Lopez L, Menendez Monzonis L, Vicente LB, Kaur J, Buschmann M. Nanouptake COST Action CA 15119. In: Report about Nanofluid's Health, Safety and Environmental Impact. Castello de la Plana (Spain). 2019. pp. 1-43

[36] Elcioglu EB, Turgut A, Murshed SMS. Socio-economic and environmental impacts of nanofluids. In: Murshed SMS, editor. Fundamentals and Transport Properties of Nanofluids. Cambridge, UK: Royal Society of Chemistry; 2022

[37] Said Z, Arora S, Bellos E. A review on performance and environmental effects of conventional and nanofluid-based thermal photovoltaics. Renewable and Sustainable Energy Reviews. 2018;**94**:302-316

[38] Nižetić S, Papadopoulos AM, Giama E. Comprehensive analysis and general economic-environmental evaluation of cooling techniques for photovoltaic panels, part I: Passive cooling techniques. Energy Conversion and Management. 2017;**149**:334-354

[39] Masuda H, Ebata A, Teramae K, Hishinuma N. Alteration of thermal conductivity and viscosity of liquid by dispersing ultra-fine particles. Dispersion of Al_2O_3, SiO_2 and TiO_2 ultra-fine particles. Netsu Bussei. 1993;7(4):227-233

[40] Eastman JA, Choi US, Li S, Thompson LJ, Lee S. Enhanced thermal conductivity through the development

of nanofluids. MRS Proceedings. 1996;**457**:3-11

[41] Eastman JA, Choi SUS, Li S, Yu W, Thompson LJ. Anomalously increased effective thermal conductivities of ethylene glycol-based nanofluids containing copper nanoparticles. Applied Physics Letters. 2001;**78**(6):718-720

[42] Siricharoenpanich A, Wiriyasart S, Naphon P. Study on the thermal dissipation performance of GPU cooling system with nanofluid as coolant. Case Studies in Thermal Engineering [Internet]. 2021;**25**:100904. DOI: 10.1016/j.csite.2021.100904

[43] Saripella SK, Yu W, Routbort JL, France DM, Uddin R. Effects of Nanofluid Coolant in a Class 8 Truck Engine. SAE Technical Paper 2007-01-2141; 2007

[44] Ijam A, Saidur R. Nanofluid as a coolant for electronic devices (cooling of electronic devices). Applied Thermal Engineering [Internet]. 2012;**32**:76-82. DOI: 10.1016/j.applthermaleng.2011.08.032

[45] Kananathan J, Samykano M, Sudhakar K, Subramaniam SR, Selavamani SK, Manoj Kumar N, et al. Nanofluid as coolant for grinding process: An overview. IOP Conference Series: Materials Science and Engineering. 2018;**342**:012078

[46] Samylingam L, Anamalai K, Kadirgama K, Samykano M, Ramasamy D, Noor MM, et al. Thermal analysis of cellulose nanocrystal-ethylene glycol nanofluid coolant. International Journal of Heat and Mass Transfer [Internet]. 2018;**127**:173-181. DOI: 10.1016/j.ijheatmasstransfer.2018.07.080

[47] Rahimi MH, Jahanfarnia G, Vosoughi N. Thermal–hydraulic analysis of nanofluids as the coolant in supercritical water reactors. The Journal of Supercritical Fluids [Internet]. 2017;**128**(May):4756. DOI: 10.1016/j.supflu.2017.05.017

[48] Genc Y, Uzun S, Acir A. The effect of nanofluid coolant and thorium-added fuel on burnup dependent isotopic compositions in VVER-1000 reactor. Afyon Kocatepe University – Journal of Science and Technology. 2024;**24**(045901):986-992

[49] Koo J, Kleinstreuer C. Impact analysis of nanoparticle motion mechanisms on the thermal conductivity of nanofluids. International Communications in Heat and Mass Transfer. 2005;**32**(9):1111-1118

[50] Iacobazzi F, Milanese M, Colangelo G, Lomascolo M, de Risi A. An explanation of the Al_2O_3 nanofluid thermal conductivity based on the phonon theory of liquid. Energy [Internet]. 2016;**116**:786-794. DOI: 10.1016/j.energy.2016.10.027

[51] Evans W, Prasher R, Fish J, Meakin P, Phelan P, Keblinski P. Effect of aggregation and interfacial thermal resistance on thermal conductivity of nanocomposites and colloidal nanofluids. International Journal of Heat and Mass Transfer. 2008;**51**(5-6):1431-1438

[52] Xuan Y, Li Q, Hu W. Aggregation structure and thermal conductivity of nanofluids. AICHE Journal. 2003;**49**(4):1038-1043

[53] Kotia A, Borkakoti S, Deval P, Ghosh SK. Review of interfacial layer's effect on thermal conductivity in nanofluid. Heat and Mass Transfer. 2017;**53**:2199-2209

[54] Domingues G, Volz S, Joulain K, Greffet JJ. Heat transfer between

two nanoparticles through near field interaction. Physical Review Letters. 2005;**94**(085901):1-4

[55] Zhao Y, Tang GH, Li ZY. Parametric investigation for suppressing near-field thermal radiation between two spherical nanoparticles. International Communications in Heat and Mass Transfer. 2012;**39**:918-922

[56] Qiu L, Zhu N, Feng Y, Michaelides EE, Żyła G, Jing D, et al. A review of recent advances in thermophysical properties at the nanoscale: From solid state to colloids. Physics Reports [Internet]. 2020;**843**:1-81. DOI: 10.1016/j.physrep.2019.12.001

[57] Turgut A, Tavman I, Chirtoc M, Schuchmann HP, Sauter C, Tavman S. Thermal conductivity and viscosity measurements of water-based TiO_2 nanofluids. International Journal of Thermophysics. 2009;**30**:1213-1226

[58] Turgut A, Sauter C, Chirtoc M, Henry JF, Tavman S, Tavman I, et al. AC hot wire measurement of thermophysical properties of nanofluids with 3ω method. European Physical Journal Special Topics. 2008;**153**:349-352

[59] Buongiorno J, Venerus DC, Prabhat N, McKrell T, Townsend J, Christianson R, et al. A benchmark study on the thermal conductivity of nanofluids. Journal of Applied Physics. 2009;**106**:094312

[60] Esfahani MA, Toghraie D. Experimental investigation for developing a new model for the thermal conductivity of silica/water-ethylene glycol (40%–60%) nanofluid at different temperatures and solid volume fractions. Journal of Molecular Liquids [Internet]. 2017;**232**:105-112. DOI: 10.1016/j.molliq.2017.02.037

[61] Xie H, Yu W, Chen W. MgO nanofluids: Higher thermal conductivity and lower viscosity among ethylene glycol-based nanofluids containing oxide nanoparticles. Journal of Experimental Nanoscience. 2010;**5**(5):463-472

[62] Hwang Y, Park HS, Lee JK, Jung WH. Thermal conductivity and lubrication characteristics of nanofluids. Current Applied Physics. 2006;**6S1**:e67-e71

[63] Darvanjooghi MHK, Nasr EM. Experimental investigation of the effect of nanoparticle size on thermal conductivity of in-situ prepared silica–ethanol nanofluid. International Communications in Heat and Mass Transfer [Internet]. 2016;**77**:148-154. DOI: 10.1016/j.icheatmasstransfer.2016.08.001

[64] Maheshwary PB, Handa CC, Nemade KR. A comprehensive study of effect of concentration, particle size and particle shape on thermal conductivity of titania/water based nanofluid. Applied Thermal Engineering [Internet]. 2017;**119**:79-88. DOI: 10.1016/j.applthermaleng.2017.03.054

[65] Turgut A, Sağlanmak Ş, Doğanay S. Nanoakışkanların Isıl İletkenlik ve Viskozitesinin Deneysel İncelenmesi: Tanecik Boyutu Etkisi (experimental investigation on thermal conductivity and viscosity of nanofluids: Particle size effect). Journal of the Faculty of Engineering and Architecture of Gazi University. 2016;**31**(1):95-103

[66] Ali FM, Yunus WMM, Talib ZA. Study of the effect of particles size and volume fraction concentration on the thermal conductivity and thermal diffusivity of Al_2O_3 nanofluids. International Journal of Physical Sciences. 2013;**8**(28):1442-1457

[67] Timofeeva EV, Routbort JL, Singh D. Particle shape effects on thermophysical

properties of alumina nanofluids. Journal of Applied Physics. 2009;**106**:014304

[68] Jeong J, Li C, Kwon Y, Lee J, Kim SH, Yun R. Particle shape effect on the viscosity and thermal conductivity of ZnO nanofluids. International Journal of Refrigeration [Internet]. 2013;**36**:2233-2241. DOI: 10.1016/j.ijrefrig.2013.07.024

[69] Huang H, Li C, Huang S, Shang Y. A sensitivity analysis on thermal conductivity of Al_2O_3-H_2O nanofluid: A case based on molecular dynamics and support vector regression method. Journal of Molecular Liquids [Internet]. 2024;**393**:123652. DOI: 10.1016/j.molliq.2023.123652

[70] Jin X, Wang R, Huang L, Shao C. Effect of interfacial layer around core-shell nanoparticles on thermal conductivity of nanofluids. Powder Technology [Internet]. 2023;**429**:118945. DOI: 10.1016/j.powtec.2023.118945

[71] Chakraborty S, Panigrahi PK. Stability of nanofluid: A review. Applied Thermal Engineering. 2020;**174**:115259

[72] Elçioğlu EB, Şimşek E, Okutucu-Özyurt T. Stability of nanofluids fundamentals, state-of-the-art, and potential applications. In: Microscale and Nanoscale Heat Transfer: Analysis, Design, and Application. Boca Raton: CRC Press; 2016. pp. 155-181

[73] Mahbubul IM, Elcioglu EB, Amalina MA, Saidur R. Stability, thermophysical properties and performance assessment of alumina – Water nanofluid with emphasis on ultrasonication and storage period. Powder Technology [Internet]. 2019;**345**:668-675. DOI: 10.1016/j.powtec.2019.01.041

[74] Wei W, Cai J, Hu X, Han Q, Liu S, Zhou Y. Fractal analysis of the effect of particle aggregation distribution on thermal conductivity of nanofluids. Physics Letters A [Internet]. 2016;**380**(37):2953-2956. DOI: 10.1016/j.physleta.2016.07.005

[75] Wang R, Qian S, Zhang Z. Investigation of the aggregation morphology of nanoparticle on the thermal conductivity of nanofluid by molecular dynamics simulations. International Journal of Heat and Mass Transfer [Internet]. 2018;**127**:1138-1146. DOI: 10.1016/j.ijheatmasstransfer.2018.08.117

[76] Cui W, Shen Z, Yang J, Wu S. Effect of chaotic movements of nanoparticles for nanofluid heat transfer augmentation by molecular dynamics simulation. Applied Thermal Engineering [Internet]. 2015;**76**:261-271. DOI: 10.1016/j.applthermaleng.2014.11.030

[77] Wang X, Jing D. Determination of thermal conductivity of interfacial layer in nanofluids by equilibrium molecular dynamics simulation. International Journal of Heat and Mass Transfer [Internet]. 2019;**128**:199-207. DOI: 10.1016/j.ijheatmasstransfer.2018.08.073

[78] Khodayari A, Fasano M, Bozorg Bigdeli M, Mohammadnejad S, Chiavazzo E, Asinari P. Effect of interfacial thermal resistance and nanolayer on estimates of effective thermal conductivity of nanofluids. Case Studies in Thermal Engineering. 2018;**12**(January):454-461

[79] Du J, Su Q, Li L, Wang R, Zhu Z. Evaluation of the influence of aggregation morphology on thermal conductivity of nanofluid by a new MPCD-MD hybrid method. International Communications in Heat and Mass Transfer [Internet]. 2021;**127**:105501. DOI: 10.1016/j.icheatmasstransfer.2021.105501

[80] Xuan Y, Li Q, Tie P. The effect of surfactants on heat transfer feature of nanofluids. Experimental Thermal and Fluid Science [Internet]. 2013;**46**:259-262. DOI: 10.1016/j.expthermflusci.2012.12.004

[81] Das PK, Islam N, Santra AK, Ganguly R. Experimental investigation of thermophysical properties of Al_2O_3–water nanofluid: Role of surfactants. Journal of Molecular Liquids [Internet]. 2017;**237**:304-312. DOI: 10.1016/j.molliq.2017.04.099

[82] Khairul MA, Shah K, Doroodchi E, Azizian R, Moghtaderi B. Effects of surfactant on stability and thermo-physical properties of metal oxide nanofluids. International Journal of Heat and Mass Transfer [Internet]. 2016;**98**:778-787. DOI: 10.1016/j.ijheatmasstransfer.2016.03.079

[83] Asadi A, Asadi M, Siahmargoi M, Asadi T, Gholami AM. The effect of surfactant and sonication time on the stability and thermal conductivity of water-based nanofluid containing $Mg(OH)_2$ nanoparticles: An experimental investigation. International Journal of Heat and Mass Transfer [Internet]. 2017;**108**:191-198. DOI: 10.1016/j.ijheatmasstransfer.2016.12.022

[84] Mahbubul IM, Shahrul IM, Khaleduzzaman SS, Saidur R, Amalina MA, Turgut A. Experimental investigation on effect of ultrasonication duration on colloidal dispersion and thermophysical properties of alumina-water nanofluid. International Journal of Heat and Mass Transfer [Internet]. 2015;**88**:73-81. DOI: 10.1016/j.ijheatmasstransfer.2015.04.048

[85] Wang XJ, Zhu DS, Yang S. Investigation of pH and SDBS on enhancement of thermal conductivity in nanofluids. Chemical Physics Letters [Internet]. 2009;**470**:107-111. DOI: 10.1016/j.cplett.2009.01.035

[86] Li XF, Zhu DS, Wang XJ, Wang N, Gao JW, Li H. Thermal conductivity enhancement dependent pH and chemical surfactant for Cu-H_2O nanofluids. Thermochimica Acta. 2008;**469**:98-103

[87] Mariano A, Pastoriza-Gallego MJ, Lugo L, Mussari L, Piñeiro MM. Co_3O_4 ethylene glycol-based nanofluids: Thermal conductivity, viscosity and high pressure density. International Journal of Heat and Mass Transfer [Internet]. 2015;**85**:54-60. DOI: 10.1016/j.ijheatmasstransfer.2015.01.061

[88] Yang B, Han ZH. Temperature-dependent thermal conductivity of nanorod-based nanofluids. Applied Physics Letters. 2006;**89**(8):1-4

[89] Agarwal R, Verma K, Agrawal NK, Singh R. Sensitivity of thermal conductivity for Al_2O_3 nanofluids. Experimental Thermal and Fluid Science [Internet]. 2017;**80**:19-26. DOI: 10.1016/j.expthermflusci.2016.08.007

[90] Das PK, Mallik AK, Ganguly R, Santra AK. Synthesis and characterization of TiO_2-water nanofluids with different surfactants. International Communications in Heat and Mass Transfer. 2016;**75**:341-348

[91] Fan W, Zhong F. Experimental study on thermal conductivity of kerosene-based nanofluids. Thermochim Acta [Internet]. 2022;**712**:179229. DOI: 10.1016/j.tca.2022.179229

[92] Chopkar I, Sudarshan S, Das PK, Manna I. Effect of particle size on thermal conductivity of nanofluid. Metallurgical and Materials Transactions A: Physical Metallurgy and Materials Science. 2008;**39**(7):1535-1542

[93] Beck MP, Yuan Y, Warrier P, Teja AS. The effect of particle size on the thermal conductivity of alumina nanofluids. Journal of Nanoparticle Research. 2009;**11**(5):1129-1136

[94] Song D, Jing D, Ma W, Zhang X. High thermal conductivity of nanoparticles not necessarily contributing more to nanofluids. Applied Physics Letters. 2018;**113**:223104

Chapter 5

Determining Thermal Conductivity Coefficient of Nanofluid by Beam Displacement Method

Soroush Javadipour, Ramin Farzadi, Faridoddin Hassani, Keyvan Homayouni, Afshin Kouhkord and Fatemeh Rezaei

Abstract

Accurate measurement of thermal properties of fluids and nanofluids is essential for optimizing performance and necessitating advanced techniques. Traditional methods, including transient and steady-state techniques, have limitations such as longer testing times and larger sample sizes. Optical methods, like beam displacement and laser interferometry, provide non-invasive, high-resolution measurements of temperature fields and thermal conductivity. The beam displacement method is an optical technique that relies on the deviation of a beam caused by changes in the refractive index of fluids due to temperature variations. Other optical method likes laser interferometry enhance accuracy and efficiency by reducing reliance on thermocouples and capturing real-time data. The advanced measurement methods provide reliable data essential for optimizing fluids and nanofluids applications in various industrial and engineering contexts. This chapter discusses different methods measuring the thermal functionality of fluids and nano-fluids with a focus on optical non-invasive methods.

Keywords: beam displacement, optical measurement, nanofluids, laser interferometry, transient measurement methods, heat transfer coefficient (HTC)

1. Introduction

Nanomaterials and nanofluids represent a significant advancement in thermal management technologies, offering unique properties that enhance heat transfer performance across various applications. Nanomaterials refer to substances or materials where at least one-dimension measures between 1 and 100 nanometers. This nanoscale range imparts unique physical and chemical properties to these materials, distinguishing them significantly from their macroscopic forms. These properties include increased surface area, enhanced reactivity, and improved mechanical strength, which make nanomaterials applicable across a broad spectrum of uses, from medicine to electronics [1–5].

Contrarily, nanofluids represent specially designed fluids that consist of a primary liquid medium, like water, oil, or ethylene glycol, with nanoparticles dispersed within.

IntechOpen

These nanoparticles, frequently crafted from metals, metal oxides, or carbon-based substances, usually fall within the size range of 1 to 100 nm. By integrating these nanoparticles into the base fluids, there is a notable improvement in thermal conductivity and heat transfer properties, positioning nanofluids as a potential enhancement for thermal efficiency across different applications [6–10].

The utilization of nanofluids spans multiple industries and applications. In the automotive sector, nanofluids are employed in vehicle cooling systems, such as radiators and engine coolants, to enhance heat dissipation and reduce the size of cooling components [11–13]. In electronics, nanofluids are used for cooling high-performance devices, including CPUs and GPUs, where efficient heat transfer is critical for maintaining performance and reliability [14]. Renewable energy systems, particularly solar thermal collectors, also benefit from nanofluids, as they improve heat absorption and overall efficiency [15–20]. Furthermore, nanofluids show promise in the realm of medicine, particularly in areas such as precise medication delivery and thermal treatments for cancer management [21–24].

The unique characteristics of nanofluids, coupled with their extensive array of uses, underscore importance of developing reliable measurement techniques to accurately assess their thermal behavior. Determining the heat transfer coefficient (HTC) of nanofluids primarily involves two main approaches: those that are transient and those that operate under steady-state conditions. Transient methods involve measuring the thermal properties of a fluid by observing its response to a sudden change in temperature or heat input. The fundamental characteristic of transient methods is their ability to capture the time-dependent behavior of temperature changes in the fluid. In these methods, a thermal disturbance is introduced, and the resulting temperature response is monitored over time. This transient response provides insights into the thermal conductivity and other thermal properties of the fluid by analyzing how quickly the fluid reaches a new thermal equilibrium. Transient methods are particularly useful for materials that exhibit rapid thermal responses due to their thermal properties.

In contrast, steady-state methods maintain a constant temperature gradient across fluid and measure the heat transfer rate once system has reached thermal equilibrium [25–28]. In this scenario, the temperature distribution remains constant over time, allowing for direct calculations of thermal conductivity and heat transfer coefficients. In contrast, steady-state methods generally necessitate extended testing periods and larger quantities of samples compared to transient methods, and they can provide more straightforward calculations and are often less sensitive to transient phenomena that could introduce variability in the results [29–31].

Transient methods include the transient hot wire method, which uses a wire which is so thin and heated by an electrical current, monitoring the temperature rise over time to calculate thermal conductivity. **Figure 1** illustrates the components necessary for the hot wire method. The laser flash method, part of the broader category of transient methods, operates by delivering a brief heat pulse to one surface of the sample. It captures the temperature increase on the sample's opposite side, leveraging the time taken for heat to spread through the material to calculate its thermal conductivity. The temperature oscillation method is another transient procedure which involves applying periodic temperature oscillations to the sample and measuring the resulting temperature change to derive thermal properties [32, 33].

Both transient and steady-state methods have their respective advantages and disadvantages. Steady-state methods are simpler to set up and analyze, directly measuring thermal properties with less complex calculations. However, they require

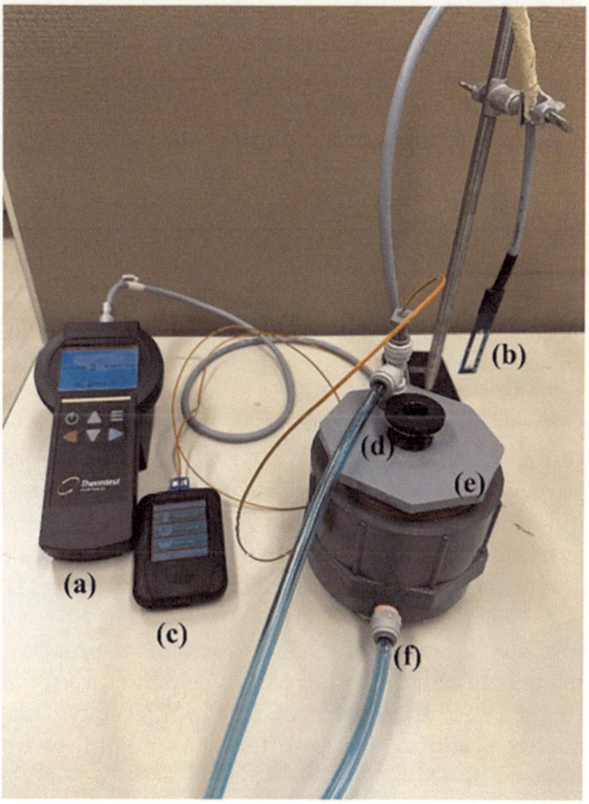

Figure 1.
Setup for the transient hot wire method, including: (a) a device for processing hot wire data, (b) a sensor for the hot wire, (c) a data logger connected to a thermocouple within a water bath, (d) a container holding the nanofluid, (e) a container filled with water, and (f) hoses leading to the water bath.

longer testing times and larger sample sizes and can be affected by heat loss. On the other hand, transient methods offer faster measurement times and require smaller sample sizes, allowing them to capture dynamic thermal behavior. However, they involve more complex calculations and can be sensitive to natural convection effects.

Optical methods represent an innovative approach for measuring the thermal properties of nano-fluids, and they can be either transient or steady state. These methods utilize light to measure thermal properties, relying on changes in either of beam path, light intensity or phase shifts due to temperature variations in nanofluid. Optical techniques can provide high-resolution, non-invasive measurements of temperature fields and thermal conductivity, making them particularly advantageous for studying nanofluids. They can capture the effects of nanoparticles on thermal properties without disturbing the fluid flow, which is critical for maintaining the integrity of the measurements [34–36].

Optical methods offer several advantages over traditional measurement techniques. They provide non-invasive measurements that do not disturb fluid flow, allowing for accurate assessments of thermal behavior. The high spatial resolution of optical techniques enables detailed visualizations of temperature fields, facilitating comprehensive analyses of thermal gradients within the fluid. Additionally, many optical methods are less sensitive to calibration errors associated with thermocouples, relying instead on optical properties, which enhances the accuracy of the results.

Furthermore, optical methods can achieve steady-state conditions more quickly than traditional techniques, improving experimental efficiency and providing rapid results that are particularly beneficial in both research and industrial applications.

Both transient and steady-state methods have their strengths and weaknesses, while optical methods present innovative solutions that address some limitations of traditional techniques. Several optical methods have been proposed for measuring, and the thermal conductivity of nanofluids is extensively explored in the subsequent sections of this discussion.

Beam displacement and laser interferometry are such techniques that use the beam deviation and interference patterns of laser light to measure temperature gradients. Temperature-induced alterations in the refractive index facilitate the estimation of thermal conductivity through the analysis of beam displacement and fringe shift changes in nanofluid.

Faris et al. [36] employed the beam displacement technique to determine HTC. Thermal pulses were supplied to the nanofluid, and the beam displacement was determined using a PID detector. In our prior work, we developed a beam displacement technique and increased the pulse period to five seconds. Increased pulse time provides increased beam displacement, which we measured using a CCD detector. Following image processing and mathematical calculations, the HTC of nanofluid was calculated [37].

The comparative interferometric method assesses the thermal conductivity of nanofluids against a benchmark fluid via a Mach–Zehnder interferometer. It achieves this by documenting the equilibrium temperature profiles in two neighboring fluids, divided by a conductive partition, thereby offering a straightforward evaluation of their thermal attributes [38–41].

2. Comparative interferometric method

The procedure for assessing the thermal conductivity of nanofluids through the comparative interferometric method involves several steps that ensure accurate and reliable results [42]. This section elaborates on the systematic approach taken during the experimental process, highlighting the setup, operational conditions, and data acquisition methods. Various sections of comparative method are illustrated in **Figure 2**.

2.1 Establishing the temperature gradient

The measurement begins by establishing a controlled temperature gradient across a thin aluminum barrier that separates two fluid chambers: one filled with the nanofluid and the other with deionized water that serves as the reference fluid. The top aluminum plate is heated using a temperature-controlled water bath, while the bottom plate is cooled by direct contact with another heat exchanger. This configuration creates a steady-state thermal condition, where the upper plate's temperature is kept around 26.5°C and the bottom plate at around 16.5°C. The resulting temperature at the interface (the separating aluminum barrier) stabilizes near room temperature, approximately 21.5°C $\pm$ 0.5°C, minimizing heat exchange with the ambient environment.

To ensure effective heat transfer and minimize measurement errors, the aluminum barrier is designed to be thin, allowing for efficient conduction while reducing

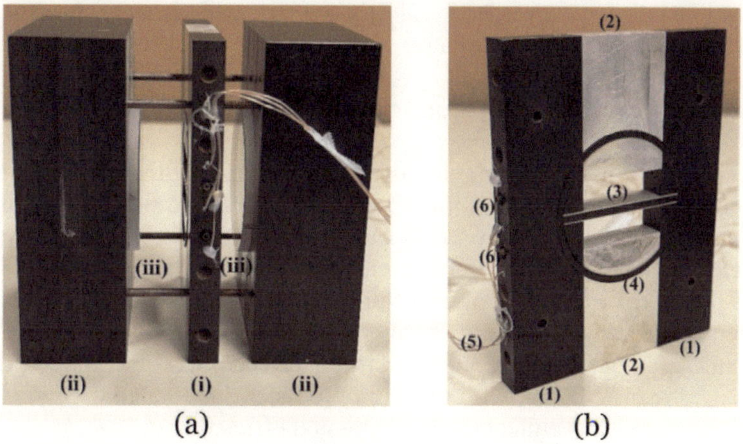

Figure 2.
Assembly of the model, comprising: (a) the central fluid compartment, ii) the two Delrin end walls, iii) the two optical windows. (b) elements of the central segment, including: (1) two Delrin slabs, (2) two aluminum plates, (3) a divider plate, (4) O-rings, (5) three thermocouples placed in their respective openings, and (6) fluid introduction ports.

thermal resistance. The setup is crucial for achieving a one-dimensional heat transfer scenario, which is necessary for applying Fourier's law in the analysis of thermal conductivity.

2.2 Data acquisition

Throughout the experiment, temperature measurements are recorded using high-precision 40-gauge T-type thermocouples are strategically positioned: one at the boundary of the dividing plate, and two others to track the internal temperatures of both the top and bottom aluminum plates' surfaces. Before conducting the experiments, these thermocouples undergo calibration in a stable temperature bath using a platinum RTD sensor, guaranteeing a precision level of $\pm 0.04°C$. This calibration is essential for minimizing uncertainties in temperature readings, which directly influence the calculated thermal conductivity.

The interferometric measurements are conducted utilizing a Mach-Zehnder interferometer, which captures temperature fields across two fluids. The optical setup allows for full-field visualization of the temperature distribution, providing a comprehensive view of the thermal behavior of the nanofluid. The interferometer operates by analyzing the phase shifts in the light waves caused by variations in refractive index of the fluids owing to temperature fluctuations. This optical method significantly reduces the reliance on thermocouple accuracy, as the primary source of measurement uncertainty arises from optical errors in fringe gradients rather than thermal measurement inaccuracies.

2.3 Achieving steady-state conditions

The system is designed to reach a steady-state condition rapidly, with measurements stabilizing within approximately five minutes. During this period, the interferograms of the temperature fields are obtained at one-minute intervals. The analysis of these interferograms allows for the determination of the temperature gradient across the aluminum barrier, which is essential for calculating the relative thermal conductivity of the nanofluid relative to the reference fluid.

2.4 The advantages of the comparative nature of the method

Independence from thermocouple calibration: Given that the optical assessments rely on modifications in refractive index instead of direct temperature readings, this approach is less susceptible to errors associated with thermocouple measurements.

Full-field visualization: The interferometric technique enables comprehensive analysis of the thermal behavior of nanofluid, allowing researchers to observe how the addition of nanoparticles affects the temperature distribution.

Rapid measurements: Compared to traditional steady-state techniques, this method is significantly faster, with the ability to reach stable measurements in a fraction of the time, thus enhancing experimental efficiency.

In summary, the measurement procedure for thermal conductivity using the comparative interferometric method is meticulously designed to ensure accurate and reliable results. By establishing a controlled temperature gradient, utilizing precise data acquisition methods, and leveraging the advantages of optical measurements, this technique provides a robust framework for investigating the thermal properties of nanofluids. The results obtained not only contribute to the understanding of nanofluid behavior but also establish a foundation for future research in thermal management applications.

3. Non-intrusive dynamic measurement

This approach leverages Mach-Zehnder interferometry to investigate aqueous colloidal suspensions of nanoparticles, specifically Al_2O_3, in deionized water [43]. The method is designed to provide high sensitivity and accuracy through whole-field measurement of temperature distribution.

The experimental process initiates with the creation of nanofluids. Aluminum oxide (Al_2O_3) nanoparticles, boasting 99.8% purity and an average size of 13 nm, are suspended in deionized water sourced from an 18.2 MΩ·cm Millipore SAS Elix 3 unit. Suspensions are formulated at concentrations of 0.02%, 0.035%, and 0.05% by mass. To achieve a uniform distribution and maintain stability for over 24 hours, these samples are mixed using an Ultra Turrex T-18 mixer operated at 8000 rpm. The size distribution and dispersion quality of the nanoparticles in the suspension are confirmed using scanning electron microscopy (SEM). Observations indicate that lower concentrations lead to small agglomerates with some chain formations, while higher concentrations result in larger agglomerates but fewer chains.

The measurement strategy utilizes Mach-Zehnder interferometry, taking advantage of the refractive index's sensitivity to temperature. This configuration uses a coherent light beam that is bifurcated into two paths: one traverses the nanofluid, and the other acts as a baseline. Upon recombination, they generate an interference pattern, or interferogram, which is continuously captured to document the changing temperature profile within the nanofluid's thermal diffusion zone. A controlled heat source induces a temperature gradient in the nanofluid contained within a test cavity. The sequence of recorded interferograms provides detailed mapping of temperature distribution across the suspension, and thermal diffusivity is calculated using the least square method applied to the temperature-position data derived from the interferograms. The non-intrusive nature of this technique ensures that measurements are free from disturbances typical of contact-based methods.

Experimental findings show a significant boost in thermal diffusivity as temperature and nanoparticle concentration rise, compared to the base fluid (deionized water). This improvement is visually substantiated through interferograms and further supported by theoretical models. Detailed investigations into microscopic processes contributing to this enhancement involve measuring ζ-potential, pH, and effective viscosity. These metrics indicate that the suspension's stability, affected by nanoparticle concentration and temperature, correlates with the formation of particle clusters, essential for boosting thermal diffusivity. Higher concentrations of negatively charged species foster stronger repulsive forces, diminishing the chance of aggregation and thus improving thermal transport capabilities.

The measured thermal diffusivity values align reasonably well with patterns observed in current research studies, reinforcing the reliability of the proposed measurement technique. The non-intrusive dynamic measurement method using Mach–Zehnder interferometry proves to be a highly effective technique for studying the thermal properties of nanofluids. It offers precise, real-time visualization and measurement of temperature distribution and thermal diffusivity. The comprehensive data obtained through this method can significantly contribute to the understanding and optimization of nanofluid applications in various high-performance cooling systems. Comparison of the interferograms obtained from the method are shown in **Figure 3**.

4. Interferometry-based inverse heat transfer approach

A suggested technique for evaluating the thermal diffusivity of diluted nanofluids through an interferometry-driven inverse heat transfer methodology involves a precise and non-intrusive technique that leverages laser interferometry to capture the transient thermal diffusion field within a test cavity [44]. Apparatuses of the presented method are illustrated in **Figure 4**. The experimental setup begins with the preparation of a rectangular test cavity, which consists of two horizontal walls (top and bottom) maintained at different temperatures to create a thermal gradient. The cavity dimensions are 25 mm along the x-axis, 20 mm along the y-axis, and 60 mm along the z-axis. The working medium, either de-ionized water or nanofluid, is situated between these two walls, which are differentially heated, with the upper wall maintained at a higher temperature compared to the lower wall. This arrangement sustains a temperature difference of 1.2 K across the fluid by employing two distinct water baths set to constant temperatures, accurate to $\pm 0.1°C$. K-type thermocouples are fixed to the surfaces of the top and bottom walls to uphold consistent temperatures, and the cavity's side walls are lined with thermocouple insulation to mitigate external temperature effects.

At the core of this measurement approach lies the application of a Mach–Zehnder interferometer to visualize the thermal landscape inside the test cavity. An aligned He-Ne laser emitting at 632.8 nm is employed, and a spatial filter generates a parallel light beam. This beam is bifurcated into two branches, the test branch and the reference branch, via a beam splitter. The test branch beam traverses the fluid within the test cavity, whereas the reference branch beam follows an equivalent path devoid of the test specimen. After reflection by mirrors, both beams converge at another beam splitter to form an interference pattern. The dynamic progression of the thermal

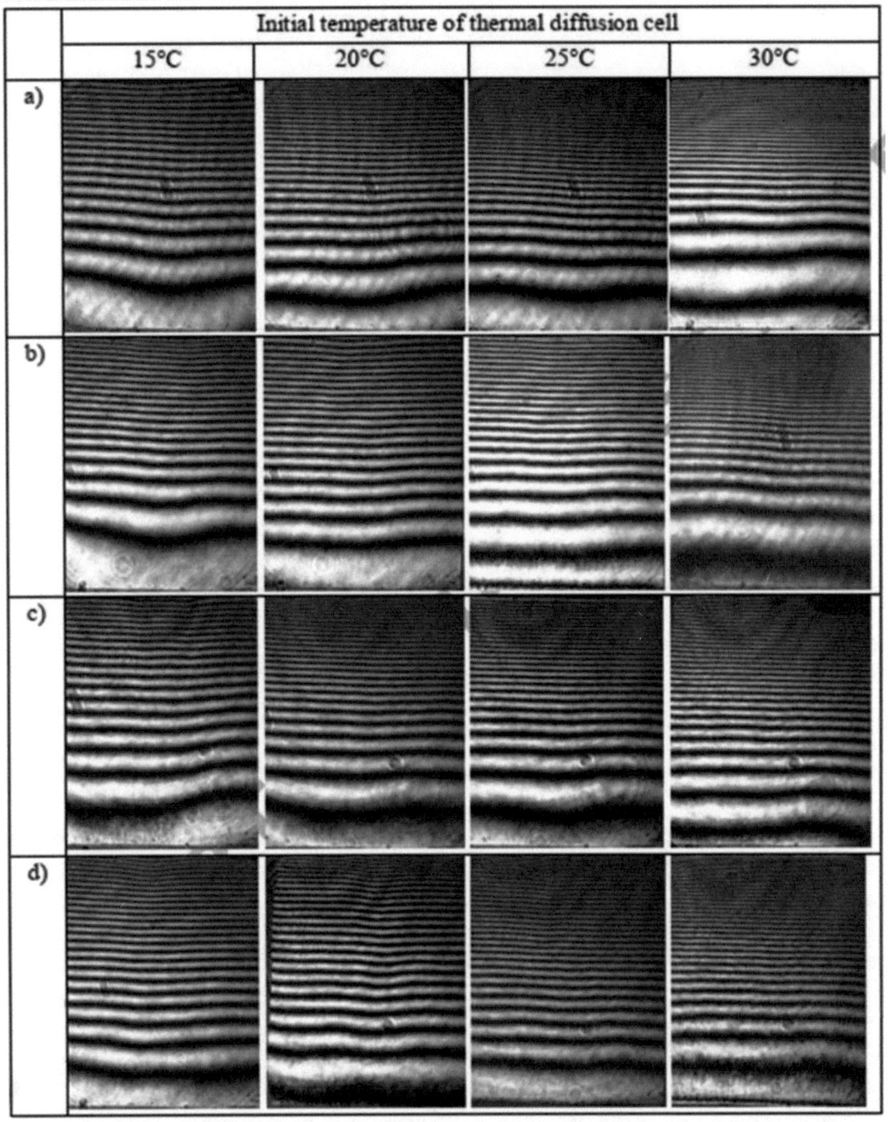

Figure 3.
Examination of interferograms captured 15 minutes post-experiment in four tests involving conductive heat transfer, conducted under similar temperature differences (ΔT = 1.5°C) but varying initial temperatures. The tested fluids include: (a) deionized water, (b) 0.02% mass Al_2O_3, (c) 0.035% mass Al_2O_3, and (d) 0.05% mass Al_2O_3 nanofluid suspensions. Each interferogram has dimensions of 50 x 40.3 mm (height x width), corresponding to the actual length scale.

diffusion field is documented through interferograms at successive time points, illustrating temperature profile across the fluid layer.

The interferometry-based method employs an inverse heat transfer technique to examine the thermal diffusivity. An analytical solution to the transient heat diffusion equation is used as a forward model for an estimated value of thermal diffusivity, with appropriate boundary conditions applied. The experimentally obtained temperature fields from the interferograms are compared with this analytical solution. By applying the concepts of inverse heat transfer, the thermal diffusivity of the working fluid is

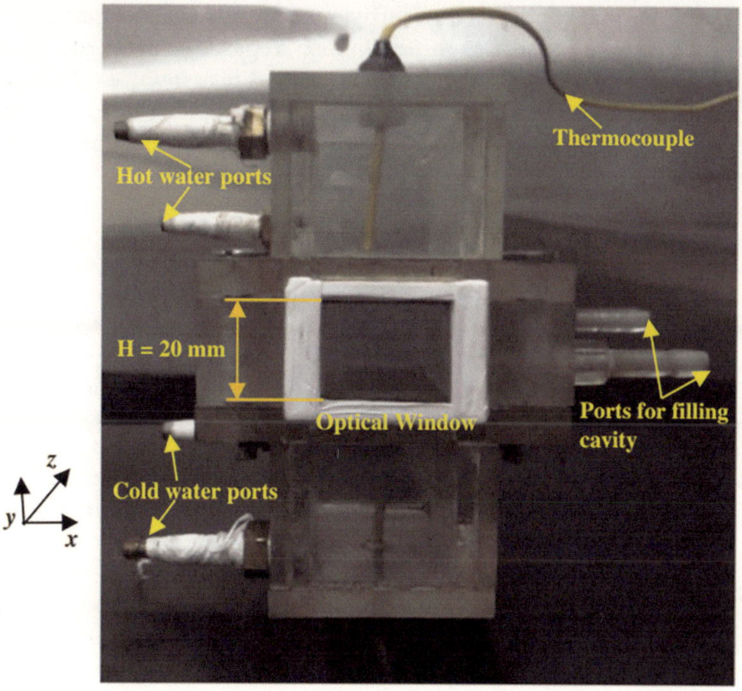

Figure 4.
Graphical depiction of a horizontally oriented test cavity with differential heating.

progressively refined to reduce the discrepancy between the experimental outcomes and those predicted analytically. The validity of the method is established by initially measuring the thermal diffusivity of the base fluid (de-ionized water) and then proceeding to evaluate the thermal diffusivity of nanofluids (SiO_2 and Al_2O_3) at different volume fractions. Findings indicate an uptick in thermal diffusivity as nanoparticle concentration rises, underscoring the method's utility.

This interferometry-based inverse heat transfer approach offers a robust and non-intrusive method for accurately examining the thermal diffusivity of dilute nanofluids. By leveraging the benefits of laser interferometry, this method overcomes the limitations associated with conventional techniques, providing high-precision thermal property measurements crucial for advancing and implementing nanofluids across various industrial and engineering applications.

5. Beam displacement method

The beam displacement method is an optical technique developed to determine the thermal conductivity coefficient of a fluids [9]. This method is based on altering the path of the beam as it passes through a fluid after experiencing a thermal shock. When a fluid is in contact with a heated surface, variations in temperature affect the velocity of fluid molecules, which in turn directly affects the density and refractive index of the fluid. As a result, the phenomena known as the thermal lens occurs. As declared, a change in the refractive index results in the deviation of the beam when it traverses a region next to a heated surface. By calculating the beam displacement, solving Eq. (10), and doing computational fluid dynamics (CFD) analysis on the liquid within the cuvette, the thermal conductivity coefficient of fluids is determined. The beam displacement technique consists of three primary steps: First, set up an optical setup to measure the displacement of the beam. The second task involves solving the computational fluid dynamics (CFD) of fluid in a cuvette. The third task involves implementing a trial-and-error loop technique to get the thermal conductivity coefficient [6].

5.1 Beam displacement optical setup

Figure 5 shows that the measurements were conducted in a vacuum chamber to minimize noise in the illustration. When particles suspended in the air traveling in the beam, they may generate noise due to scattering and blocking the light, so it is better to perform this experiment in vacuum or clean box. A He-Ne laser emitting at 632 nm wavelength and producing 2 mW of output power was utilized. Also, the polarizer and analyzer were responsible for regulating the intensity of the laser beam. To further narrow the beam width, two pinholes with hole diameters of 200 and 50 microns were placed in the path of the beam. Additionally, the XYZ stage performed the alignment, and a specially made cap on the cuvette fixed the heater in the cuvette.

A CCD detector recorded the beam displacement, and the computer received the data. As shown in the picture, the temperature in nearby of the heater is greater than the temperature at other points; hence, the beam path ought to pass in close proximity to the heater.

5.2 Beam displacement calculation

Faris et al. [9] drive this equation in their studies, as seen below:

$$\theta_{nf} = \frac{1}{n}\frac{dn}{dT}\int_{-W}^{W}\frac{D}{\sqrt{D^2+Z^2}}\frac{dT}{d\rho}dz \tag{1}$$

In this equation, n is the nanofluid's refractive index, and T is the temperature. D, Z, and ρ represent the distance between the heater and the probe beam, and the

Figure 5.
Beam displacement experimental setup.

adjacent side of a triangle with ρ as the hypotenuse in the picture. Also, we could be driving Eq. (3) from Snell's low, as seen below:

$$n_f \sin\theta_f = n_{air} \sin\theta_{air} \tag{2}$$

If we consider n_{air} is equal to 1, and $\sin\theta \cong \theta$, thus:

$$\theta_{air} = n_f \theta_f \tag{3}$$

Beam displacement (δ_{Totall}) consists of beam deviation in fluid (δ_f) and beam deviation in air (δ_{air}).

$$\delta_{Totall} = \delta_f + \delta_{air} \tag{4}$$

and also:

$$\delta_f = \theta_f (W - w) \tag{5}$$

$$\delta_{air} = \theta_{air} L_{air} \tag{6}$$

In the previous equations, W is the inner diameter of the cuvette, w is half of it, and L_{air} is the distance between the cuvette and the CCD detector. If Eq. (3) is merged into Eq. (7), then:

$$\delta_{air} = n_f \theta_f L_{air} \tag{7}$$

So,

$$\delta_{Totall} = ((W - w) + n_f L_{air})\theta_f \tag{8}$$

By merging Eq. (1) and Eq. (9) and considering the term of time, we have:

$$\delta_{Totall=} \frac{W - w + n\, L_{air}}{n} \frac{dn}{dT} \int_0^t \int_0^W \frac{D}{\sqrt{D^2 + Z^2}} \frac{dT}{d\rho} dz dt \tag{9}$$

5.3 Beam displacement numerical simulation

The second step in the beam displacement techniques included simulating variations in temperature inside the cuvette that contained the heater. The finite element method was used to computationally examine the temperature evolutions in a numerical model. The actual dimensions of the experimental setup, as seen in **Figure 6b**, were used to simulate the geometry of the cuvette containing the fluid and cylindrical heater. Furthermore, the extremely fine mesh type used in the solution is depicted in **Figure 6a**. The cuvette and cylindrical heater were taken to be solid objects with certain densities, thermal conductivities, and heat capacities to specify the material characteristics. For the external borders, we used thermal insulation conditions, and we considered the heater to be a heat source with an 80 MW/m^3 heat generating power.

Furthermore, the physical problem with the interface, which was time-dependent, was solved by using the heat transfer equation. We collected temperature data for all

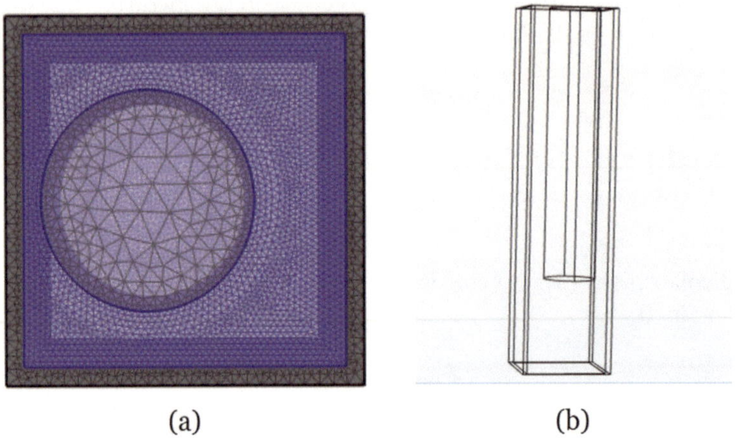

Figure 6.
(a) is applying mesh on geometry of problem, (b) geometry of problem based on actual dimension of experimental setup.

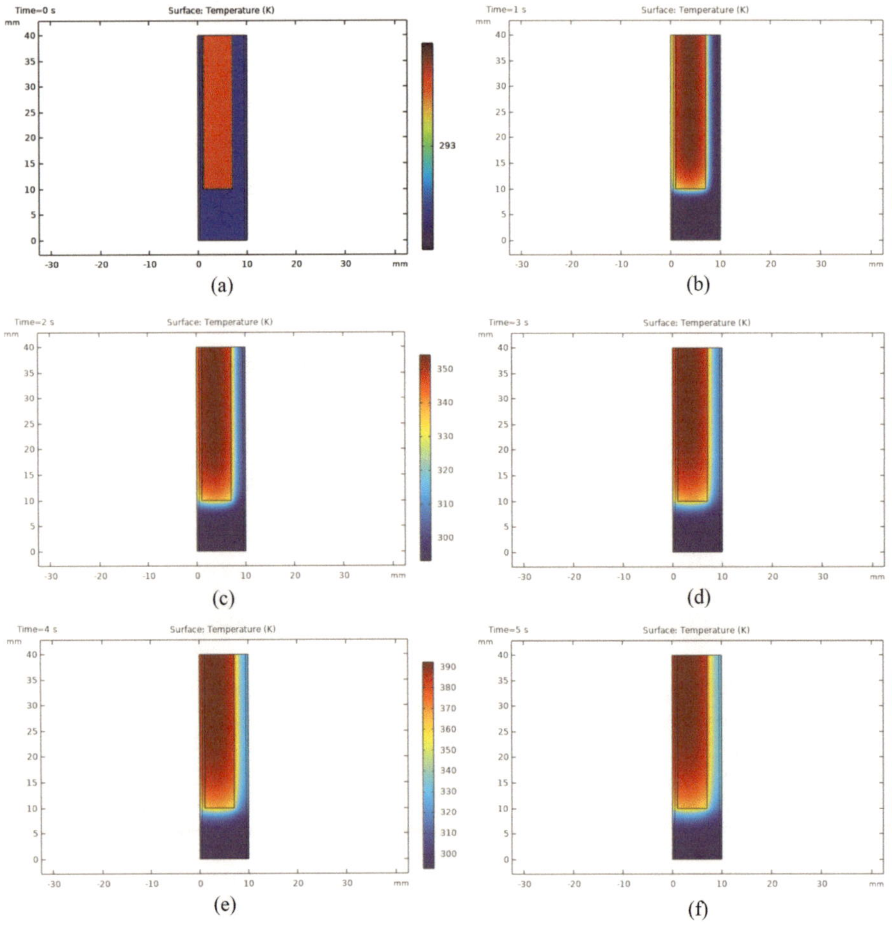

Figure 7.
Result of numerical simulation in different time, (a): time = 0 s, (b): time = 1 s, (c): time = 2 s, (d): time = 3 s, (e): time = 4 s, (f): time = 5 s.

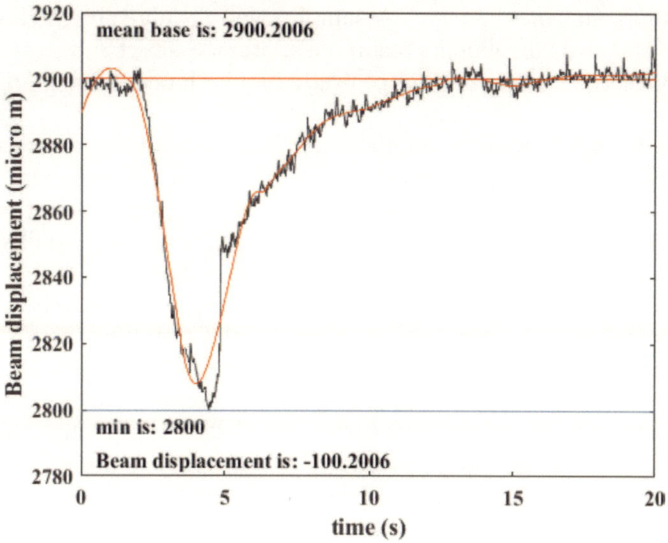

Figure 8.
Result of experimental evaluation.

areas during a period of 5 seconds. To compute the term of $\frac{dT}{d\rho}$ in Eq. (10), we measure the temperature throughout the whole beam path in the cuvette and identify the line that is about 400 micrometers from the heater.

5.4 HTC of water by beam displacement method

The numerical simulation returned a value of 100. 2 micrometers for DI water, with a heat transfer coefficient of 5.6 w.m/k. **Figure 7** shows the numerical simulation results of DI water. Over time, the temperature of the water around the heater increases, leading to the occurrence of thermal lens phenomena and displacement of the beam. The beam displacement of DI water is depicted in **Figure 8**. Experimental evaluation was conducted in 5 seconds applying the thermal shock. The inner size of cuvette was 9 mm, and $\frac{dT}{d\rho}$ and n are 0.000156 and 1.33 for DI water, respectively. The distance between surface of cuvette and CCD detector was 19.5 mm approximately. As seen in the figure, the diagram is denoised using MATLAB's smooth data tool. The average baseline was calculated by calculating the mean value before and after the heat shock. The result showed that the mean was 2900.2 and the pick was 2800, so the beam displacement was 100.2 micrometer that have a suitable coincidence with numerical simulation.

6. Conclusions

Accurate measurement of the thermal properties of nanofluids is critical for understanding and optimizing their performance in practical applications. Techniques for measuring the heat transfer coefficient (HTC) of nanofluids can be broadly categorized into transient and steady-state methods, each with its own advantages and limitations. Transient methods, such as the transient hot wire and laser flash methods, offer the benefit of capturing the dynamic thermal behavior of fluids, requiring shorter measurement times and smaller sample sizes. Steady-state methods, like the parallel plates and coaxial cylinders techniques, provide direct and straightforward calculations of thermal properties but typically require longer testing times and larger samples.

Optical methods, including beam displacement, laser interferometry and Mach–Zehnder interferometry, present innovative solutions that address many limitations of traditional techniques. These methods offer non-invasive, high-resolution measurements of temperature fields and thermal conductivity, providing detailed visualizations of thermal gradients without disturbing the fluid flow.

The comparative interferometric method, which utilizes a controlled temperature gradient across a thin aluminum barrier separating the nanofluid from a reference fluid, has demonstrated significant advantages in measuring thermal conductivity. This method's reliance on optical measurements reduces the sensitivity to thermocouple inaccuracies and provides comprehensive visualizations of temperature distribution, enhancing the accuracy and efficiency of the measurements.

Similarly, the non-intrusive dynamic measurement method using Mach–Zehnder interferometry offers a robust approach for studying thermal diffusivity. This technique provides precise, real-time visualization of temperature distribution and thermal diffusivity, contributing valuable data for understanding and optimizing nanofluid applications.

The HTC of DI water was determined by beam displacement method. The result shows that 100.2 micrometer deviation in beam path during applying 5 second thermal shock. Beam displacement represents another advanced method for measuring the thermal conductivity coefficient of dilute nanofluids. With development in more exact detection of beam displacement using more improved detectors in the future and improving in image processing by AI development, HTC of fluids may be determined precisely.

In conclusion, the development and refinement of measurement techniques for the thermal properties of nanofluids are essential for advancing their application in thermal management systems. The innovative use of optical methods and interferometry has opened new avenues for precise, non-intrusive measurements, providing critical insights into the thermal behavior of nanofluids. These advancements not only enhance our understanding of nanofluids but also pave the way for their optimized use in real-world applications, contributing to more efficient and effective thermal management solutions across various industries.

Author details

Soroush Javadipour[1*], Ramin Farzadi[2], Faridoddin Hassani[3], Keyvan Homayouni[4], Afshin Kouhkord[5] and Fatemeh Rezaei[6]

1 Faculty of Materials Science and Engineering, K. N. Toosi University of Technology, Tehran, Iran

2 Mechanical Engineering Department, K. N. Toosi University of Technology, Tehran, Iran

3 Department of Mechanical Engineering, University of Mohaghegh Ardabili, Ardabil, Iran

4 Department of Petroleum Engineering, Science and Research Branch, Islamic Azad University, Tehran, Iran

5 Department of Mechanical Engineering, K. N. Toosi University of Technology, Tehran, Iran

6 Department of Physics, K. N. Toosi University of Technology, Tehran, Iran

*Address all correspondence to: sorooshjavadipoor@gmail.com

References

[1] Bokov D, Turki Jalil A, Chupradit S, Suksatan W, Javed Ansari M, Shewael IH, et al. Nanomaterial by sol-gel method: Synthesis and application. Advances in Materials Science and Engineering. 2021;**2021**:1-21. Available from: https://www.hindawi.com/journals/amse/2021/5102014/

[2] Sanjay ST, Zhou W, Dou M, Tavakoli H, Ma L, Xu F, et al. Recent advances of controlled drug delivery using microfluidic platforms. Advanced Drug Delivery Reviews. 2018;**128**:3-28

[3] Kouhkord A, Hassani F, Amirmahani M, Golshani A, Naserifar N, Moghanlou FS, et al. Controllable microfluidic system through intelligent framework: Data-driven modeling, machine learning energy analysis, comparative multiobjective optimization, and experimental study. Industrial and Engineering Chemistry Research. 2024;**63**:13326-13344. DOI: 10.1021/acs.iecr.4c00456

[4] Hassani F, Kouhkord A, Golshani A, Amirmahani M, Sadegh Moghanlou F, Naserifar N, et al. Micro-electro-mechanical acoustofluidic mixing system: A response surface-metaheuristic machine learning fusion framework. Expert Systems with Applications. 2024;**249**:123638. Available from: https://linkinghub.elsevier.com/retrieve/pii/S0957417424005037

[5] Shafiq M, Anjum S, Hano C, Anjum I, Abbasi BH. An overview of the applications of nanomaterials and nanodevices in the food industry. Food. 2020;**9**(2):148. Available from: https://www.mdpi.com/2304-8158/9/2/148

[6] Javadipour S, Shokuhfar A, Heidary Z, Amiri Roshkhar MA, Homayouni K, Rezaei F, et al. Stability, optimum ultrasonication, and thermal and electrical conductivity estimation in low concentrations of Al12Mg17 nanofluid by dynamic light scattering and beam displacement method. Scientific Reports. 2023;**13**(1):13659. Available from: https://www.nature.com/articles/s41598-023-40844-9

[7] Hassani F, Sadegh Moghanlou F, Minaei A, Vajdi M, Golshani A, Kouhkord A, et al. An efficient framework for controllable micromixer design through the fusion of data-driven modeling and machine learning insights: Numerical and experimental analysis. Physics of Fluids. 2024;**36**(3):033606. Available from: https://pubs.aip.org/pof/article/36/3/033606/3268733/An-efficient-framework-for-controllable-micromixer

[8] Makabenta JMV, Nabawy A, Li CH, Schmidt-Malan S, Patel R, Rotello VM. Nanomaterial-based therapeutics for antibiotic-resistant bacterial infections. Nature Reviews. Microbiology. 2021;**19**(1):23-36. Available from: https://www.nature.com/articles/s41579-020-0420-1

[9] Ali FM, Yunus WMM, Moksin MM, Talib ZA. The effect of volume fraction concentration on the thermal conductivity and thermal diffusivity of nanofluids: Numerical and experimental. The Review of Scientific Instruments. 2010;**81**(7):0749011-0749019. Available from: https://pubs.aip.org/rsi/article/81/7/074901/924337/The-effect-of-volume-fraction-concentration-on-the

[10] Kouhkord A, Amirmahani M, Hassani F, Naserifar N. Machine learning and metaheuristics in microfluidic transport characterization and optimization: CFD and experimental

study integrated with predictive modelling. Canadian Journal of Chemical Engineering. 2024. DOI: 10.1002/cjce.25430

[11] Bencs P, Alktranee M. The potential of vehicle cooling systems. Journal of Physics Conference Series. 2021;**1935**(1): 012012. DOI: 10.1088/1742-6596/1935/1/012012

[12] Xian HW, Sidik NAC, Saidur R. Hybrid nanocoolant for enhanced heat transfer performance in vehicle cooling system. International Communications in Heat and Mass Transfer. 2022;**133**: 105922. Available from: https://linkinghub.elsevier.com/retrieve/pii/S0735193322000446

[13] Kılınç F, Buyruk E, Karabulut K. Experimental investigation of cooling performance with graphene based nano-fluids in a vehicle radiator. Heat and Mass Transfer. 2020;**56**(2):521-530. DOI: 10.1007/s00231-019-02722-x

[14] Siricharoenpanich A, Wiriyasart S, Naphon P. Study on the thermal dissipation performance of GPU cooling system with nanofluid as coolant. Case Studies in Thermal Engineering. 2021;**25**: 100904. Available from: https://linkinghub.elsevier.com/retrieve/pii/S2214157X21000678

[15] Hatamleh RI, Rawa MJH, Abu-Hamdeh NH, Shboul B, Karimipour A. Simulation of nanofluid flow in a solar panel cooling system to investigate the panel's electrical-thermal efficiency with artificial neural network. Journal of the Taiwan Institute of Chemical Engineers. 2023;**148**:104879. Available from: https://linkinghub.elsevier.com/retrieve/pii/S1876107023002080

[16] Tian MW, Khetib Y, Yan SR, Rawa M, Sharifpur M, Cheraghian G, et al. Energy, exergy and economics study of a solar/thermal panel cooled by nanofluid. Case Studies in Thermal Engineering. 2021;**28**:101481. Available from: https://linkinghub.elsevier.com/retrieve/pii/S2214157X21006444

[17] Rahmani M, Shahabi Nejad A, Fallah Barzoki M, Kasaeian A, Sameti M. Simulation of solar absorption refrigeration cycle with CuO nanofluid for summer cooling of a residential building. Thermal Science and Engineering Progress. 2022;**34**:101419. Available from: https://linkinghub.elsevier.com/retrieve/pii/S2451904922002256

[18] Khalili Z, Sheikholeslami M. Investigation of innovative cooling system for photovoltaic solar unit in existence of thermoelectric layer utilizing hybrid nanomaterial and Y-shaped fins. Sustainable Cities and Society. 2023;**93**:104543. Available from: https://linkinghub.elsevier.com/retrieve/pii/S2210670723001543

[19] Zarei A, Izadpanah E, Babaie RM. Using a nanofluid-based photovoltaic thermal (PVT) collector and eco-friendly refrigerant for solar compression cooling system. Journal of Thermal Analysis and Calorimetry. 2023;**148**(5):2041-2055. DOI: 10.1007/s10973-022-11850-2

[20] Tong Y, Ham J, Cho H. Investigation of thermo-optical properties and photothermal conversion performance of MWCNT, Fe_3O_4, and ATO nanofluid for volumetric absorption solar collector. Applied Thermal Engineering. 2024;**246**: 123005. Available from: https://linkinghub.elsevier.com/retrieve/pii/S1359431124006732

[21] Mannu R, Karthikeyan V, Velu N, Arumugam C, Roy VAL, Gopalan AI, et al. Polyethylene glycol coated

magnetic nanoparticles: Hybrid nanofluid formulation, properties and drug delivery prospects. Nanomaterials. 2021;**11**(2):440. Available from: https://www.mdpi.com/2079-4991/11/2/440

[22] Alnahdi AS, Nasir S, Gul T. Blood-based ternary hybrid nanofluid flow-through perforated capillary for the applications of drug delivery. Waves in Random and Complex Media. 2022;**32**: 1-19. DOI: 10.1080/17455030.2022.2134607

[23] Shahzadi I, Bilal S. A significant role of permeability on blood flow for hybrid nanofluid through bifurcated stenosed artery: Drug delivery application. Computer Methods and Programs in Biomedicine. 2020;**187**:105248. Available from: https://linkinghub.elsevier.com/retrieve/pii/S0169260719318589

[24] Sheikhpour M, Arabi M, Kasaeian A, Rokn Rabei A, Taherian Z. Role of nanofluids in drug delivery and biomedical technology: Methods and applications. Nanotechnology, Science and Applications. 2020;**13**:47-59. Available from: https://www.dovepress.com/role-of-nanofluids-in-drug-delivery-and-biomedical-technology-methods–peer-reviewed-article-NSA

[25] Loong TT, Salleh H. A review on measurement techniques of apparent thermal conductivity of nanofluids. IOP Conference Series Materials Science and Engineering. 2017;**226**:012146. DOI: 10.1088/1757-899X/226/1/012146

[26] Xu G, Fu J, Dong B, Quan Y, Song G. A novel method to measure thermal conductivity of nanofluids. International Journal of Heat and Mass Transfer. 2019; **130**:978-988. Available from: https://linkinghub.elsevier.com/retrieve/pii/S0017931018332605

[27] Kouhkord A, Ghanbarzadeh A, Ebrahimi P, Najafi E. Design of a genetic based optimized fuzzy logic controller for enhanced trajectory tracking accuracy of a 3P robot. In: 10th RSI International Conference on Robotics and Mechatronics (ICRoM) 2022, Iran, Tehran. IEEE; 2022

[28] Golshani A, Kouhkord A, Ghanbarzadeh A, Najafi E. Control design for safe human-robot collaboration based on ISO/TS 15066 with power and force limit. In: 2023 11th RSI International Conference on Robotics and Mechatronics (ICRoM), Iran, Tehran. IEEE; 2023. pp. 279-284. Available from: https://ieeexplore.ieee.org/document/10412570/

[29] Paul G, Chopkar M, Manna I, Das PK. Techniques for measuring the thermal conductivity of nanofluids: A review. Renewable and Sustainable Energy Reviews. 2010;**14**(7):1913-1924. Available from: https://linkinghub.elsevier.com/retrieve/pii/S1364032110000729

[30] Yiamsawasd T, Dalkilic AS, Wongwises S. Measurement of the thermal conductivity of titania and alumina nanofluids. Thermochimica Acta. 2012;**545**:48-56. Available from: https://linkinghub.elsevier.com/retrieve/pii/S0040603112003127

[31] Ahammed N, Asirvatham LG, Titus J, Bose JR, Wongwises S. Measurement of thermal conductivity of graphene–water nanofluid at below and above ambient temperatures. International Communications in Heat and Mass Transfer. 2016;**70**:66-74. Available from: https://linkinghub.elsevier.com/retrieve/pii/S073519331500233X

[32] Cabaleiro D, Colla L, Agresti F, Lugo L, Fedele L. Transport properties and heat transfer coefficients of ZnO/(ethylene glycol + water) nanofluids.

International Journal of Heat and Mass Transfer. 2015;**89**:433-443. Available from: https://linkinghub.elsevier.com/retrieve/pii/S001793101500561X

[33] Ma B, Kumar N, Kuchibhotla A, Banerjee D. Estimation of measurement uncertainties for thermal conductivity of nanofluids using transient plane source (TPS) technique. In: 2018 17th IEEE Intersociety Conference on Thermal and Thermomechanical Phenomena in Electronic Systems (ITherm). IEEE; 2018. pp. 178-186. Available from: https://ieeexplore.ieee.org/document/8419622/

[34] Srivastava A, Muralidhar K, Panigrahi PK. Optical imaging and three dimensional reconstruction of the concentration field around a crystal growing from an aqueous solution: A review. Progress in Crystal Growth and Characterization of Materials. 2012;**58**(4):209-278. Available from: https://linkinghub.elsevier.com/retrieve/pii/S096089741200023X

[35] Berto A, Mattiuzzo N, Zanetti E, Meneghetti M, Del Col D. Measurements of solar energy absorption in a solar collector using carbon nanofluids. Renewable Energy. 2024;**230**:120763. Available from: https://linkinghub.elsevier.com/retrieve/pii/S0960148124008310

[36] Faris Mohammed A, Mat Yunus WM, Talib ZA. Study of the effect of particles size and volume fraction concentration on the thermal conductivity and thermal diffusivity of Al_2O_3 nanofluids. International Journal of Physical Science. 2013;**28**(8):1442-1457

[37] Javadipour S, Shokuhfar A, Homayouni K, Heidary Z, Rezaei F. Investigation into the size distribution of Al_2O_3- ZnO nanoparticles dispersed in DI water and following the impact of CNTs on the stability, heat transfer, and electricity transfer of Al_2O_3- ZnO-CNT hybrid nano. Defect and Diffusion Forum. 2022;**420**:172-192. Available from: https://www.scientific.net/DDF.420.172

[38] Ji Y, Chung Y, Sprinzak D, Heiblum M, Mahalu D, Shtrikman H. An electronic Mach–Zehnder interferometer. Nature. 2003;**422**(6930):415-418. Available from: https://www.nature.com/articles/nature01503

[39] Busch P, Shilladay C. Complementarity and uncertainty in Mach–Zehnder interferometry and beyond. Physics Reports. 2006;**435**(1):1-31. Available from: https://linkinghub.elsevier.com/retrieve/pii/S0370157306003322

[40] Dandridge A. Fiber optic sensors based on the Mach–Zehnder and Michelson interferometers. In: Fiber Optic Sensors. United States: Wiley; 2024. pp. 213-248. DOI: 10.1002/9781119678892.ch10

[41] Sreekumar S, Ganguly A, Khalil S, Chakrabarti S, Hewitt N, Mondol JD, et al. Thermo-optical characterization of novel MXene/carbon-dot hybrid nanofluid for heat transfer applications. Journal of Cleaner Production. 2024;**434**:140395. Available from: https://linkinghub.elsevier.com/retrieve/pii/S0959652623045535

[42] Sahamifar S, Naylor D, Yousefi T, Friedman J. Measurement of the thermal conductivity of nanofluids using a comparative interferometric method. International Journal of Thermal Sciences. 2024;**199**:108890. Available from: https://linkinghub.elsevier.com/retrieve/pii/S1290072924000127

[43] Nimdeo YM, Srivastava A. Understanding the temperature dependence of thermo-physical properties of nanofluid suspensions using non-intrusive dynamic measurements. Experimental Thermal and Fluid Science. 2018;**94**:109-121. Available from: https://linkinghub.elsevier.com/retrieve/pii/S0894177718300463

[44] Rao SS, Srivastava A. Measuring thermal diffusivity of dilute nanofluids using interferometry-based inverse heat transfer approach. Journal of Thermophysics and Heat Transfer. 2020; **34**(3):476-487. DOI: 10.2514/1.T5826